Crystallization and Applications

Edited by Youssef Ben Smida and Riadh Marzouki

Published in London, United Kingdom

IntechOpen

Supporting open minds since 2005

Crystallization and Applications
http://dx.doi.org/10.5772/intechopen.92520
Edited by Youssef Ben Smida and Riadh Marzouki

Contributors
Atef Korchef, Maite Del Pilar Rada-Mendoza, José Luis Arciniegas Herrera, Patricia Vélez Varela, Mazen Alshaaer, Youssef Ben Smida, Riadh Marzouki, Mosbah Jemmali, Basma Marzougui, Mohamed Triki, Shinichi Sakurai, Amit Kumar Pandey, Ivana Kuta Smatanova, Barbora Kascakova, Petra Havlickova, Tatyana Prudnikova, Josefina Pons, Francisco Sánchez-Férez, Endar Hidayat, Hiroyuki Harada, Khalil Issa, Abdulaziz Alanazi, Khalid A. Aldhafeeri, Ola Alamer

First published in London, United Kingdom, 2022 by IntechOpen
IntechOpen is the global imprint of INTECHOPEN LIMITED, registered in England and Wales, registration number: 11086078, 5 Princes Gate Court, London, SW7 2QJ, United Kingdom
Printed in Croatia

British Library Cataloguing-in-Publication Data
A catalogue record for this book is available from the British Library

Additional hard and PDF copies can be obtained from orders@intechopen.com

Crystallization and Applications
Edited by Youssef Ben Smida and Riadh Marzouki
p. cm.
Print ISBN 978-1-83969-316-8
Online ISBN 978-1-83969-317-5
eBook (PDF) ISBN 978-1-83969-318-2

For EU product safety concerns:
IN TECH d.o.o., Prolaz Marije Krucifikse Kozulić 3, 51000 Rijeka, Croatia, info@intechopen.com or visit our website at intechopen.com.

Meet the editors

Youssef Ben Smida obtained a Ph.D. in Chemistry at the Faculty of Sciences of Tunis, University of Tunis El Manar, Tunisia, in 2015. He completed a postdoc at the Materials and Crystal Chemistry Laboratory, Faculty of Sciences of Tunis, where he was responsible for the X-ray diffractometers. Since June 2017, he has held a permanent position as a young researcher at the National Research Center in Materials Science, Parke Technology of Borj Cedria Tunisia. He works on energy (Na-ion batteries and fuel cells) and the environment (photocatalysis). His other areas of interest include computational methods (DFT and MD) and crystallography. Dr. Smida has co-supervised and supervised master's degree students, end-of-study engineering projects, and theses.

Riadh Marzouki is an Assistant Professor in International Cooperation between the Department of Chemistry, Faculty of Sciences, King Khalid University, Saudi Arabia, and the Department of Chemistry, Faculty of Sciences, University of Sfax, Tunisia. He is also a researcher in the Laboratory of Materials and Crystallography, University of Tunis El Manar, Tunisia. His research activities include investigating the synthesis, crystal structure, electrical properties, and biological activities of new crystalline materials. He is also the author and the co-author of more than sixty scientific papers, four book chapters, and four book editions.

Contents

Preface

Solid materials can be crystalline, amorphous, or glassy. A crystalline material is a solid whose constituents (atoms, molecules, or ions) are assembled in a regular pattern. Crystallization is a chemical reaction that makes it possible to obtain crystalline materials. Crystallized materials have three forms: single crystal, polycrystalline powder, and thin film. The phenomenon that allows crystals to grow to dimensions greater than a few millimeters is called crystal growth.

The preparation of a single crystal is the best way to obtain new materials [1]. However, the search for new phases following this path is not easy, especially for non-oxide materials. In addition, the parameters to be controlled during the synthesis differ from one method to another. For example, it was found that the solid-solid method is the simplest method to implement, but most of the parameters in this type of reaction are uncontrollable. However, in the hydrothermal method, it is possible to control the concentration, pH, solution/solvent ratio, temperature, and pressure. The crystals are initially characterized by their sizes, colors, and morphologies. Small dimensions and twins are the problems most encountered during the experiment and of course have known phases.

To obtain new phases, researchers can change the stoichiometric proportions of the reactants and the controllable parameters of the reaction process. In recent years, it has become possible to predict the existence of a crystal structure using the Universal Structure Predictor: Evolutionary Xtallography (USPEX) code [2]. Currently, this code is used to predict the existence of new phases in two-element chemical systems and some three-element chemical systems. It is currently undergoing further development to accommodate the study of larger systems; this new code is known as COPEX [3] (co-evolutionary crystal structure prediction algorithm for complex systems). USPEX has made it possible to predict the possibility of obtaining new phases at high pressure such as NaCl3 and NaCl7 [4]. In these types of materials, the inner core electrons participate in the formation of chemical bonds. Other predicted materials showed superconductivity properties at temperatures close to room temperature [5, 6]. With this code, the experimental researcher can reduce the number of experiments. In addition, it is possible to proceed directly to the synthesis of polycrystalline powder or thin-film, once crystallographic data of the new phase is obtained.

The crystallization process may be used in chemistry, physics, or materials science to prepare materials for special applications such as batteries, fuel cells, and optics. Crystallized materials may be prepared in the laboratory as powder or thin film and in some cases as a single crystal. In industry, crystallization is an elementary process necessary to obtain some pure products. The crystallization process is used also in biology to crystallize protein and in pharmacology to prepare some drugs. The crystal growth of materials is a technique used to generate material in an orderly arrangement of atoms, especially when the material shows good properties worthy of further studies such as optical properties. In this case, the growth of the materials in crystalline form allows for the elimination of defects and improvement of material properties.

Youssef Ben Smida
National Centre of Research in Materials Science,
Technology Park of Borj Cedria,
Borj Cedria, Tunisia

Riadh Marzouki
King Khalid University,
Abha, Saudi Arabia

References

[1] Y.B. Smida, R. Marzouki, S. Kaya, S. Erkan, M.F. Zid, A.H. Hamzaoui, Synthesis methods in solid-state chemistry, Synthesis Methods and Crystallization, IntechOpen2020.

[2] C.W. Glass, A.R. Oganov, N. Hansen, USPEX—Evolutionary crystal structure prediction, Computer physics communications, 175 (2006) 713-720.

[3] X. Liu, H. Niu, A.R. Oganov, COPEX: co-evolutionary crystal structure prediction algorithm for complex systems, npj Computational Materials, 7 (2021) 1-11.

[4] A. Oganov, Discovering new materials and new phenomena with evolutionary algorithms, APS March Meeting Abstracts, 2016, pp. F20. 005.

[5] A.G. Kvashnin, D.V. Semenok, I.A. Kruglov, I.A. Wrona, A.R. Oganov, High-temperature superconductivity in a Th–H system under pressure conditions, ACS applied materials & interfaces, 10 (2018) 43809-43816.

[6] I.A. Kruglov, D.V. Semenok, H. Song, R. Szczęśniak, I.A. Wrona, R. Akashi, M.M.D. Esfahani, D. Duan, T. Cui, A.G. Kvashnin, Superconductivity of LaH 10 and LaH 16 polyhydrides, Physical Review B, 101 (2020) 024508.

Section 1

Crystallization in the Inorganic Chemistry

Chapter 1

Polycrystalline Powder Synthesis Methods

Mosbah Jemmali, Basma Marzougui, Youssef Ben Smida, Riadh Marzouki and Mohamed Triki

Abstract

The synthesis of polycrystalline powder is a key step for materials sciences. In this chapter, we present the well-known methods of preparation of powders such as: solid-state reaction, sol–gel, hydrothermal, combustion, co-precipitation. Moreover, synthesis methods by arc furnace, by heating in a "high frequency" induction furnace and by high energy grinding are presented. The obtained powders could be defined by their purity, gain size, crystallinity, and morphology, which are influenced by the synthesis method. In addition, each method is dependent on some parameters like pH, concentration and temperature.

Keywords: solid-state, sol–gel, hydrothermal, combustion, coprecipitation, intermetallic

1. Introduction

Optimizing the physico-chemical properties of materials is a challenge for many scientific researchers, and it could be altered using different synthesis methods. The synthesis of ceramics is influenced by many parameters; the choice of such a synthesis method is generally conditioned by the morphology as well as by the particle size and the specific surface of the samples, which it is desired to synthesis. Ceramics, when prepared by different routes, exhibit different properties, even with the same starting compositions. There are two ways of synthesizing ceramics:

- Dry method (solid–solid method)
- Wet method (sol–gel, hydrothermal, combustion and coprecipitation methods)

Microstructure plays an important role in the properties of materials. In fact, for ceramic materials, the presence of porosities between the grains has a negative effect on the conductive properties. In addition, ceramic materials made of small grain size are thermally more stable mechanically than samples with larger grain size [1]. For example, barium hexaferrite ($BaFe_{12}O_{19}$) could be used either as a permanent magnet or as a recording medium depending on the morphology of the compound, which depends on the route of preparation [2].

A variety of metal oxides, both simple and complex, is prepared by conventional ceramic process. This involves the mixing of metal oxides, carbonates, and their repeated heating and grinding. This method is used on both a laboratory and industrial scale. Nevertheless, the need for alternative routes to the synthesis of oxides has arisen due to intrinsic problems related to:

- Inhomogeneity of the products obtained by ceramic methods
- Incorporation of chemical impurities during repeated grinding and heating operations. Impurities have an undesirable effect on physicochemical behavior
- Particle size obtained by conventional means is not fine, which makes them unsuitable

To have better control of stoichiometry, structure, and desired phase purity, mild chemical pathways are becoming increasingly important in preparing a variety of oxides, including nanocrystalline oxide materials [3].

This chapter is divided in two parts. The first one is dedicated to powders prepared by chemical methods while the second describes the methods of preparation of intermetallic compounds.

2. Powders prepared by chemical methods

2.1 Solid-state method

It is the simplest and most widely used synthesis method in solid-state chemistry. It consists of making the various solid reagents react directly at high temperature [4]. The reagents used for this method are simple oxides: ZnO, Cr_2O_3, Fe_2O_3, NiO, MgO, fluorides: BaF_2, CaF_2 [5], carbonates: $NaCO_3$, K_2CO_3 [6], nitrates: $CuNO_3$ [7] and oxalates: $Fe(COO)_2$. There are reagents, which are difficult to handle, due to their hygroscopy and high sensitivity to water vapor. For example, lanthanum oxide (La_2O_3), reacts quickly to the air to form carbonates and hydroxides, so it should be calcined first at 900 °C to remove all traces of impurities. The powders are weighed according to the stoichiometric quantities provided by the reaction equation and should grounded in an agate mortar to obtain the desired phase. It is done in the absence or in the presence of a volatile liquid (acetone) to ensure good dispersion of the grains and good homogeneity of the powder [8]. Then, the materials are subjected to a controlled heat treatment, would to form well-defined crystalline structures. The powder obtained could be annealed as it is or shaped into pellets pressed uniaxially. During calcination, there is release of carbon dioxide or oxygen dioxide and possibly a little water vapor. Successive anneals after intermediate grinding are generally necessary for the reaction to be complete. Synthesis temperatures are generally of the order of 900 to 1000 °C.

Although this method is considered the simplest and most economical compared to other methods, it has drawbacks. In fact, the reagents should be of high purity, because some impurities could have marked effects and do not give the desired phase and therefore, it would influence the properties of materials. In addition, this method leads to the formation of powders with large particles and with low homogeneity. Finally, the synthesis of ceramics by this method requires reactions for hours or even days at relatively high temperatures.

The materials synthesized by this method are numerous, among them, the cuprates A_2BO_4: La_2CuO_4, Pr_2CuO_4, Sr_2CuO_4) [9], the perovskites ABO_3: ($CaTiO_3$, $LaTiO_3$, $BaTiO_3$ [10], the pyrochlore $A_2B_2O_7$: $Dy_2Ti_2O_7$, $La_2Zr_2O_7$, $Cd_2Re_2O_7$ [11], the brownmillerite $A_2B_2O_5$: $Ca_2(Al,Fe)_2O_5$ [12].

2.2 Combustion method

The combustion method is a synthetic route for ceramics that requires less calcination compared to the conventional solid–solid method. It is simple, economical, and fast. It is interesting from research on the mechanism of the physicochemical processes involved, the dynamics of product formation, their limit of stability and process control.

Indeed, the method of combustion or synthesis by fire is known under the name of self-propagating synthesis at high temperature. To generate fire, fuel and temperature are needed. All these three elements form a triangle of fire. Fire could be described as uncontrolled combustion, which produces heat, light, and ash. The process uses highly exothermic redox chemical reactions between an oxidant and a fuel [13].

For the reaction to propagate on its own, the heat released should be greater than the heat required to initiate combustion. A redox reaction involves simultaneous oxidation and reduction processes. The classic definition of oxidation is the addition of oxygen or any other electronegative (non-metallic) element, while reduction is the addition of hydrogen or any other electropositive element (metal). The general formula of the combustion reaction is as follows:

$$\text{Combustible} + O_2 \rightarrow CO_2(g) + H_2O(g) \quad (1)$$

These reactions are highly exothermic or even explosive. This method is used for the synthesis of refractory materials: borides, nitrides, oxides. Reagents are generally used in the form of nitrates because the nitrate group acts as an oxidizing agent and it has high solubility and allows homogenization of greater product yields. They are mixed, then they are "ignited" by laser, electric arc, or heating resistance. Materials based on hydrazine, urea, glycine, or citric acid form the fuel for this reaction [14, 15].

The combustion method was unexpectedly discovered during the reaction between aluminum nitrate and urea [16, 17]. A mixture of $Al(NO_3)_3.9H_2O$ and a urea solution, when rapidly heated to about 500 °C in a muffle furnace, to burn with an incandescent flame producing a bulky white product, which has been identified as α-Al_2O_3.

2.3 Sol–gel method

Sol–gel process is considered as a soft chemistry method used for the synthesis of ceramics [18]. The sol–gel process is a wet-chemical technique that uses either a chemical solution or colloidal particles (sol) to produce an integrated network (gel). This later is a semi-rigid solid where the solvent is held captive in the network of solid material, which could be colloidal (concentrated sol) or polymeric [19].

The elaboration of materials by the sol–gel process, takes place via reactions of inorganic polymerizations in solution from molecular precursors, generally metal alkoxides: $M(OR)_n$ where M is a metal of oxidation degree n (for example: Si, Ti, Zr, Al, Sn...) and OR an alkoxyl group. After a drying step, a heat treatment leads to the elimination of organic compounds to form the inorganic oxide material. These reactions lead to gelation and the formation of a gel made up of M-OM (or M-OH-M) chains, the viscosity of which increases over time. This gel still contains solvents and precursors, which have not reacted. The "gel" phase in the sol–gel process is defined and characterized by a solid 3D "skeleton" embedded in liquid phase. The solid phase is typically a condensed polymeric sol where the particles have become entangled to form a three-dimensional network. The reactions allowing this material to be obtained are carried out at room temperature (**Figure 1**) [20, 21].

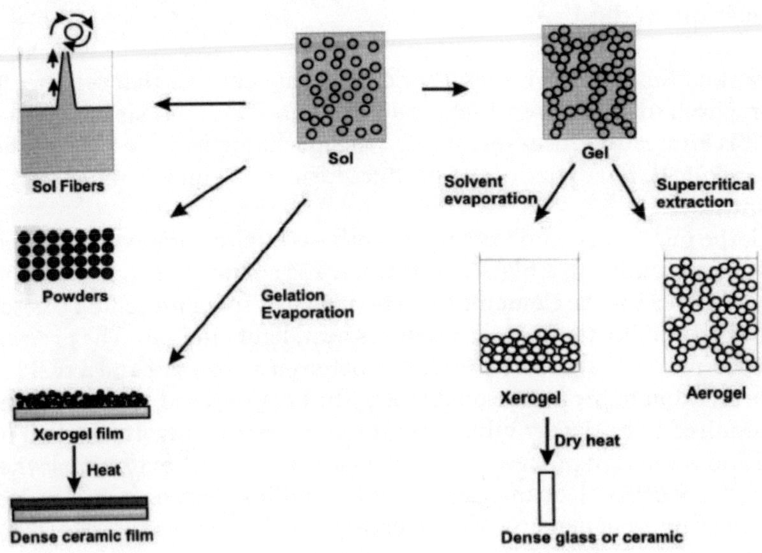

Figure 1.
General scheme of sol–gel process.

The parameters influencing the reactions are the temperature, the pH, the nature of the precursor and the solvent and the concentrations of the reactants. However, the most significant are the pH and the $[H_2O]$ / [M] ratio. The pH influences the morphology of the gel formed. In fact, an acidic pH accelerates hydrolysis and slows down condensation unlike basic pH. A high rate of hydrolysis (acidic pH) therefore promotes network growth and leads to a polymeric solution. Under acid catalysis, which is the fastest synthesis route, the gel formed is called a "polymer gel": after gelling, an open structure is obtained. However, a low rate of hydrolysis (basic pH) rather favors nucleation and leads to the formation of a colloidal solution. In the case of basic catalysis, the size of the pores is controllable (unlike acid catalysis). The gel formed is called a "colloidal gel" and has a large pore structure (clusters) [22]. Once gelation takes place, the material undergoes drying due to capillary forces in the pores and this drying could result in volume shrinkage. The drying process for obtaining the sol–gel material requires that the alcohol or water could escape as the gel solidifies. The evaporation process occurs through existing holes and channels in the porous sol–gel material. The method of drying dictate whether an aerogel or xerogel is formed. In fact, aerogel is obtained when the liquid phase of a gel is replaced by a gas in such a way that its solid network is retained, with only a slight or no shrinkage in the gel. It was usually achieved under supercritical conditions of the solvent. It is characterized by low density and high porosity. When the drying of gel is occurred by simple evaporation, xerogel is obtained. It may retain its original shape, but often cracks due to the extreme shrinkage that is experienced while being dried. After a drying process, the liquid phase is removed from the gel. Then, a thermal treatment (calcination) may be performed in order to favor further polycondensation and enhance mechanical properties [23, 24].

Although this process requires the use of relatively expensive precursors, it has many advantages:

- The high purity and the homogeneity of the solutions linked to the fact that the different constituents are mixed at the molecular scale in solution

- The controls of the porosity of the materials and the size of the nanoparticles
- The heat treatments required at low temperatures as well as the synthesis of materials inaccessible by other techniques

Several materials were synthesized by the sol–gel process. CeO_2-TiO_2 mixed oxides aerogels with high surface area and stable anatase phase are obtained via sol–gel process [25, 26]. Bismuth barium ferrite nanoparticles doped with sodium and potassium metal ions ($Bi_{0.85-x}MXBa_{0.15}FeO_3$, x = 0 and 0.1, M = Na and K) were prepared by the sol–gel method by Haruna et al. [27]. Ahmoum et al. [28] synthesized $Zn_{1-x}Ni_xO$ by sol–gel method. $La_2Mo_2O_9$ nanoparticles were synthesized by Zhang et al. [29] by the sol–gel process using lanthanum nitrate $La(NO)_3.6H_2O$, and ammonium heptamolybdate $(NH_4)_6Mo_7O_{24}.4H_2O$ as precursors.

2.4 Coprecipitation method

Another classic soft chemistry synthesis method commonly developed in the chemical industry is coprecipitation. This method involves bringing into play several metal cations and a precipitating agent, which is most commonly ammonium bicarbonate (NH_4HCO_3) or ammonium hydroxide (NH_4OH). The use of ammonium bicarbonate results in powders, which are less agglomerated than with ammonium hydroxide or urea [30].

The coprecipitation method could be performed by:

- Direct way: addition of a solution of precipitating agent in the solution of metal salts
- Reverse way: adding the metal salt solution to the precipitating agent solution

Synthesis by coprecipitation allows obtaining homogeneous ceramics and a well-controlled particle size. Precipitation is characterized by two main processes: nucleation, the birth of crystals, and crystal growth. These two phenomena determine the size distribution of the crystals. Indeed, nucleation corresponds to the formation of the smallest crystalline entity called "germ" or "nucleus". Once supersaturation is established (supersaturation is the driving force behind crystallization processes), it takes a certain time, called the induction time, for germs to appear. Crystal growth corresponds to the spontaneous evolution of germs of critical size towards a state of greater stability [31].

The coprecipitation method offers some advantages:

- Fast and easy preparation
- Easy control of particle size and composition
- Various possibilities to modify the surface state of the particles and the overall homogeneity
- Low temperature
- Does not involve the use of organic solvents

However, it has some disadvantages:

- Not applicable to uncharged cash
- Traces of impurities could be precipitated with the product
- This method does not work well if the reagents have different precipitation rates

For example, the synthesis of the $Ce_{0.5}Fe_{0.5}O_2$, cerium and iron nitrates taken in stoichiometric proportions are mixed with stirring. The basic precipitation medium consists of an aqueous solution of ammonia. The precursors are then precipitated dropwise from the reaction medium with vigorous stirring. The solution is left to mature for 2 hours at room temperature before being washed with an aqueous ammonia solution. The powder, strongly hydrated, is then placed in an oven overnight. After a grinding step, a final calcination step at 600 °C for 5 hours, to remove traces of nitrates, would be necessary [32].

2.5 Hydrothermal method

In a sealed vessel (bomb, autoclave), solvents can be brought to temperatures well above their boiling points by the increase in autogenous pressures resulting from heating. Performing a chemical reaction under such conditions is referred to as solvothermal processing or, in the case of water as solvent, hydrothermal processing.

Hydrothermal reactors are mostly metal autoclaves with Teflon or alloy linings or containing an extra beaker or tube made of Teflon, platinum, or silver to protect the autoclave body from the highly corrosive solvent, which is held at high temperature and pressure.

The influence of the conditions of hydrothermal synthesis (pH, temperature, the nature of the precursors or the presence of surfactant) on the morphology and size of the particles has been investigated by several researchers. Indeed, a high pH favors the formation of nanoparticles. On the other hand, a lower pH favors the formation of nanoparticles, which is explained by the evolution of the nature and the proportion of soluble species in solution. The increase in temperature causes an increase in particle size with a change in morphology [33, 34]. The synergistic effect of high temperature and pressure provides a one-step process to produce high crystalline materials without the need of post-annealing treatments. The hydrothermal method becomes useful when it is difficult to dissolve the precursors at low temperatures or room temperature.it prove to be useful to grow nanoparticles if the material has a high vapor pressure near its melting point or crystalline phases are not stable at melting point [35].

In addition, the size of the particles of the precursor is an important parameter in controlling the morphologies obtained. Nanotubes resulting from the hydrothermal synthesis of particles of small grain size (8–10 nm) have been proven to have a larger outer diameter (10–30 nm), thicker layers and an ill-defined tubular shape or incomplete. In contrast, large particles (~ 200 nm) lead to the formation, not of nanotubes, but rather of layers [36].

The main disadvantage of hydrothermal method is the high cost of equipment. However, it has many advantages; it could generate nanomaterials, which are stable at elevated temperatures. In addition, it creates larger sized and high-quality crystals and nanoparticles. In addition, this method could be combined with other process like microwave, electrochemistry, ultrasound, optical radiation and hot

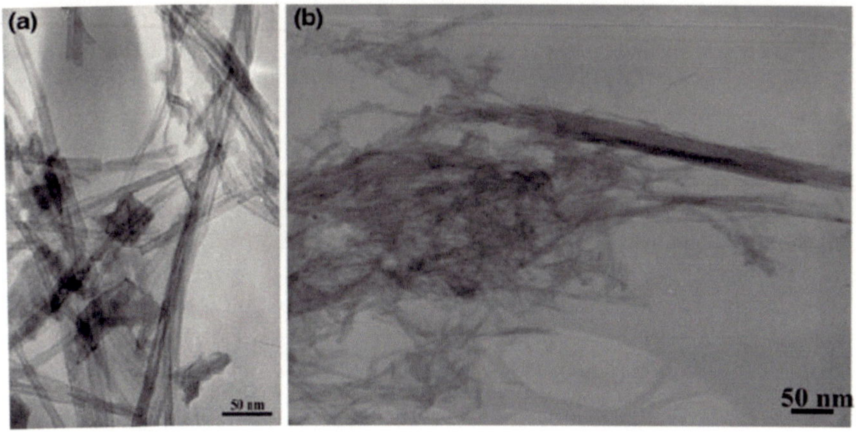

Figure 2.
TEM images of hydrogenotitanate prepared by hydrothermal method (a) nanotubes and (b) nanowires.

pressing. By varying the synthesis conditions (t, T p, etc.) it is possible to vary phase composition and particle size and to change the morphology. Hydrogenotitanate nanotubes and nanowires (**Figure 2**), used as supports for ruthenium catalysts, were obtained by hydrothermal method using highly concentrated NaOH and KOH, respectively [37].

3. Preparation of intermetallic compounds

3.1 Synthesis by arc furnace

The synthesis of the intermetallic compounds was carried out either by induction melting, by electric arc melting or by grinding, from a mixture of pure elements (purity 99.99%). This mixture contains the elements (R: Rare Earth, M: Transition Metal, T: Metalloids) in an amount corresponding to the desired stoichiometry, except in the case of Sm-based compounds for which an excess of Sm (4% - proportion mass) to have a stable phase.

The arc furnace uses the thermal energy of the electric arc established between the tungsten electrode (W) and the metals placed in a copper crucible to obtain a sufficient temperature to melt them (**Figure 3**). The technologies used can be the cold crucible (crucible formed of cooled copper or stainless-steel sectors), the hot

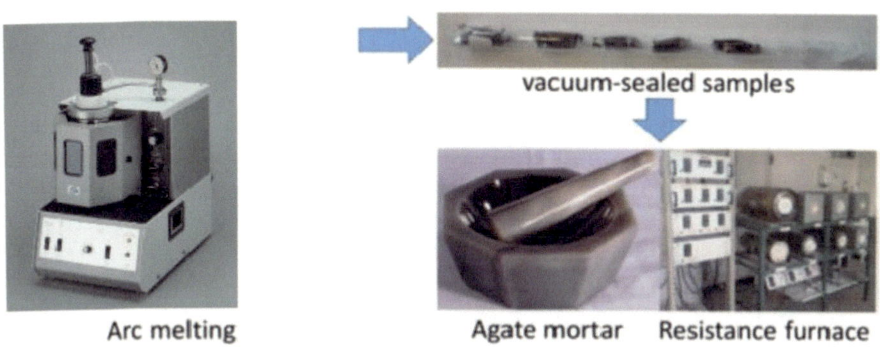

Figure 3.
Synthesis steps using arc furnace method.

crucible (graphite or others) or the use of a susceptor or muffle to radiate on the material to be heated. As a result, cold electrical insulating materials (ceramic, glass, enamels, silicon) can be treated in our induction furnaces. Plasma solutions can be considered as a means of purification of materials. The tungsten electrode and the base copper are connected to a generator that delivers a high intensity current creating an electric arc that could reach temperatures above 3000 °C.

The principle is to create an electric arc between the tungsten tip of a welding machine (Genesis 160, Imax = 160 A) and the surface of the sample. Care should be taken not to touch the sample to prevent it from sticking to the tip. The electric arc created induces a rise in temperature and the fusion of the various constituents. The crucible is cooled by circulating cold water, under high purity argon gas. Zirconium-titanium alloy was used as an O_2 getter during the melting process, which allows for a sudden quenching when the electric arc is stopped. In order to ensure good homogenization, it is necessary to move the tip to the surface of the sample but also to perform several fusions by inverting the ingot between each heating. Care should be taken to work in a slight argon overpressure and to perform several air-argon rinses in order to purify the atmosphere of the bell. Systematic weighing after reaction makes it possible to assess the loss of mass of the most volatile elements; this is related to the purity of the final products and limits to a certain extent (≤1%) the appreciation of the areas of homogeneity.

The arc furnace used could hold approximately 5 g of sample. For larger masses, the use of an induction furnace is necessary. Annealing is a homogenization technique that involves putting the sample at a high temperature for a specified time. Prolonged heating causes an increase in thermal agitation and diffusion coefficients, which allows atoms to organize themselves better, to find an optimal structure corresponding to thermodynamic equilibrium at this temperature.

The furnace used for the annealing is controlled by a Microcor Coreci regulator and the thermocouple is platinum/platinum rhodium (± 5 °C). The bulk was wrapped in a tantalum sheet. Samples were sealed in vacuum quartz tubes and annealed at 800 °C for one week in order to reach a good homogenization and improve the atomic diffusion kinetics. This temperature was chosen as a good compromise between relatively fast diffusion kinetics and absence of reaction with the quartz tube (**Figure 3**).

Intermetallic single crystals of compositions, $Er_6Fe_{17.66}Al_{5.34}Co_{0.65}$ [38], $GdFe_{0.37}Ge_2$, $GdFe_{0.27}Ge_2$ [39], $ErFe_{2.4}Al_{0.6}$ [40] and polycrystalline $SmNi_2$ [41], $SmNi_5$ [42], $Gd_2Fe_{17-x}Cu_x$ [43], $Nd_2Fe_{17-x}Co_x$ [44], $Er_6Fe_{23-x}Al_x$ [45], $SmNi_2Fe$ [46], $SmNi_{3-x}Fe_x$ (x = 0; 0,3 and 0,8) [47], $GdFe_{12-x}Cr_x$ [48] were prepared by arc furnace. These iron-based ternary diagrams R (Sm, Gd, Er, Nd)–Fe- M (Al, Cr, Cu, Co, Ni) were achieved using arc furnace method [49–56]. The stoichiometric composition of $SmNi_5$ is confirmed by SEM and XRD analyses (**Figure 4**).

3.2 Synthesis by heating in a "high frequency" induction furnace

The induction furnace or high frequency furnace (**Figure 5**) consists of an external coil inside which a non-inductive copper crucible, divided into sectors and cooled by circulating water, supports the sample.

Prolonged annealing at high temperature under secondary vacuum between 1200 and 1750 °C should be carried out for two reasons:

- To complete the reaction (synthesis)
- To obtain crystals (crystallogenesis)

DOI: http://dx.doi.org/10.5772/intechopen.97006

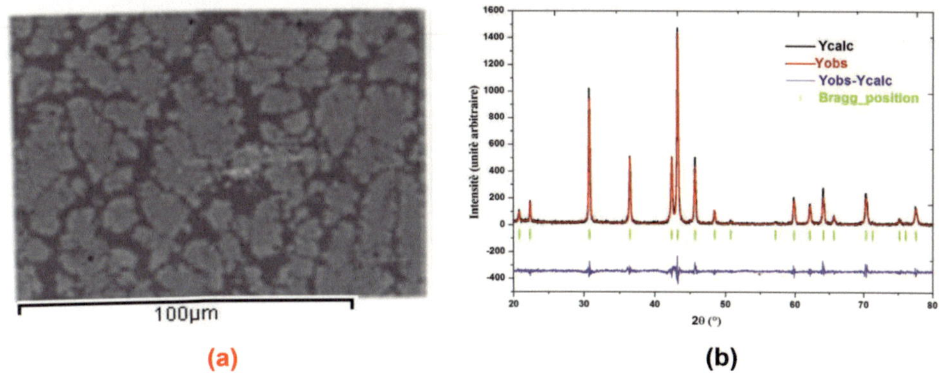

Figure 4.
$SmNi_5$ characterization: (a) SEM image and (b) XRD pattern and the corresponding Rietveld refinement.

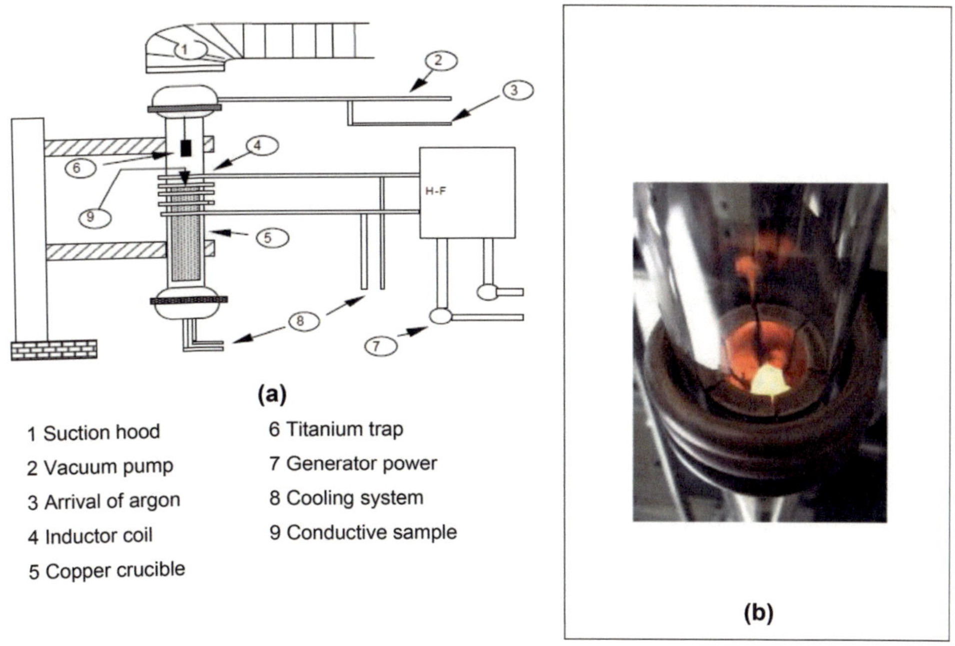

Figure 5.
(a) Diagram of the induction furnace (b) photograph of the induction furnace.

The coil is traversed by a high frequency current (from 10 to 100 kHz), which generates a variation of the magnetic field. Therefore, the metal sample, located in the center of the coil, is subjected to an induced current or eddy current, which causes the metal to heat up by Joule. The sample preparation is identical to that for the arc furnace preparation. The pure metals are placed in the copper crucible.

As the temperature rises, each sector of the latter being cooled with water, any contamination metal through the crucible is avoided. The crucible is protected from the outside environment by a glass tube in which a secondary vacuum is created before melting. This makes it possible to get rid of all gaseous species adsorbed on the internal walls of the tube and on the surface of the sample. The power of the high frequency generator is then gradually increased until all the metals melted to combine with each other. The evaporation of metals is controlled with the manometer indicating the vacuum state. Because of a significant increase in pressure is observed, argon is introduced into the tube. Five fuses in total are performed with successive turning of the alloy between each of them to ensure the homogeneity of the sample.

The induction furnace could synthesize up to 12 g of massive compounds. Unlike the arc furnace, the melting is less violent because the temperature is controllable. However, the preparation time in the induction furnace is longer because to return the sample between each fusion, it is necessary to disassemble the glass tube and therefore make the secondary vacuum again later to proceed with the next fusion.

The polycrystalline $PrTiFe_{11-x}Co_x$ ($x \leq 3$) [57] and $YFe_{11-x}Co_xTiC$ alloys (x = 0; 0.5; 1; 1.5, 2) [58]. Ingots were prepared by induction melting under argon atmosphere.

3.3 Synthesis by high energy grinding

To carry out the fragmentation of the particles, it is necessary to set them in motion in suitable equipment. Several studies [59, 60] have been carried out on the grinding of polymers using vibrating fragmentation systems. Even if this technique gives interesting results, this type of device is difficult to extrapolate industrially because of the technological difficulties linked to the vibrating system. Thus, the two crushers used were ball mill and agitated ball mill. These two mills have the same operating principles but the energy transmitted to the powder to effect the fragmentation is different. We would see later that some processes require several tens of hours in the ball mill while they only require a few tens of minutes in the agitated ball mill.

For preparation by mechanosynthesis, the starting materials should be in powder form. This poses a problem due to their sensitivity to oxidation, especially in the case of rare earths, which are very reactive to air. By means of the mechanical composition, it is possible to manufacture amorphous alloys, supersaturated solid solutions of immiscible elements in thermodynamic equilibrium, semi-crystalline compounds, as well as unregulated metal alloys.

The "Pulverisette 4" planetary vario-mill [61] is a mill consisting of an animated plate, with a rotational movement on which two jars are placed; rotating around their axis in both directions relative to that of the movement without rotation of

Figure 6.
Planetary mill Fritsch (a) planetary mill "Pulverisette 4"; (b) jar and grinding balls; (c) movement of the balls inside the jars.

the plate. The ratio between the speed of the jars ω and the plate speed Ω is called the multiplicity factor, if $\Omega/\omega > 1$ the grinding is in shock mode, and if $\Omega/\omega < 1$ we say that the grinding is in friction. As the directions of rotation of the disc and the jars are opposite, the centrifugal forces resulting from these movements act on the contents of the jars by creating high-energy effects of shocks, friction of the balls on the walls of the jars in all directions and cause the powder to fission. Grinding is carried out discontinuously; it is interrupted every 30 minutes for 15 minutes, to limit the temperature rise inside the jars, as well as to avoid the problems of powder clogging on the walls of the jars, which prevents further crushing. Previous experimental studies have identified the necessary parameters; to avoid increasing the temperature inside the jars, as well as to obtain a satisfactory nanometric powder. The polycrystalline $LaFe_{13-x}Si_x$ (x = 1.4, 1.6, 1.8, 2.0) compounds were synthesized by high energy ball milling using LaSi as a precursor to prevent oxidation of lanthanum (**Figure 6**) [62].

4. Conclusion

The synthesis of new crystalline materials is a field of intense activity in solid-state chemistry. In this chapter, the most adopted methods of synthesis of the crystalline materials are described. The first part is dedicated to solid-state method, combustion reaction, hydrothermal method, sol–gel process and coprecitation method. These methods lead to different types of crystalline phase such as organic and hybrids compounds, organometallics, oxides, intermetallic alloys, etc. However, a control of several parameters such as temperature, pressure, etc. is necessary.

The second part is devoted to intermetallic compounds synthesized by arc furnace, induction furnace and high energy grinding methods.

The arc furnace is a system using electrothermal energy provided by an electric arc to heat a load. Electric arc furnaces are used almost exclusively for the manufacture of steel from scrap recovery or arcs develop between each of three electrodes in graphite and metal, this last melt by direct attack bows digging wells in the solid metal charge.

The induction furnace is an electric furnace operating thanks to the phenomenon of metal induction heating. This method presents several advantages; it is clean, energy efficient and allows better control of the metals melting. The capacities of induction furnaces vary from less than one kilogram to one hundred tons; they are used to melt iron and steel, copper, aluminum and precious metals.

Mechanosynthesis is a very high energy milling technique, which allows to obtain metastable crystalline or nanocrystalline phases and to transform crystalline phases into amorphous phases.

Author details

Mosbah Jemmali[1,2], Basma Marzougui[3], Youssef Ben Smida[3]*, Riadh Marzouki[4,5] and Mohamed Triki[3]

1 Faculty of Science, University of Sfax, Sfax, Tunisia

2 Department of Chemistry, College of Science and Arts, Ar-Rass, Qassim University, Saudi Arabia

3 Laboratory of Valorization of Useful Materials, Techno Park of Borj Cedria, National Centre of Research in Materials Sciences, University of Carthage, Soliman, Tunisia

4 Chemistry Department, College of Science, King Khalid University, Abha, Saudi Arabia

5 Faculty of Sciences of Tunis, Laboratory of Materials, Crystal Chemistry and Applied Thermodynamics, University of Tunis El Manar, Tunisia

*Address all correspondence to: youssef_smida@yahoo.fr

References

[1] Aruna ST, Mukasyan AS. Combustion synthesis and nanomaterials. Current Opinion in Solid State and Materials Science. 2008;**12**:44

[2] Vadivelan S, Victor Jaya N. Investigation of magnetic and structural properties of copper substituted barium ferrite powder particles via co-precipitation method. Results in Physics. 2016;**6**:843

[3] Varma A, Rogachev AS, Mukasyan AS, Hwang S. Combustion Synthesis of Advanced Materials: Principles and Applications. Advances in Chemical Engineering. 1998;**24**:79

[4] Smart L, Moore E. Introduction à la Chimie du Solide. Paris: Masson; 1997

[5] Patro LN, Hariharan K. Fast fluoride ion conducting materials in solid state ionics: An overview. Solid State Ionics. 2013;**239**:41

[6] Lu Z, Yu J, Wu J, Effat MB, Stephen CT, Kwok SCT et al. Enabling room-temperature solid-state lithium-metal batteries with fluoroethylene carbonate-modified plastic crystal interlayers. Energy Storage Materials. 2019;**18**:311

[7] Rao GJ, Mazumder R, Bhattacharyya S, Chaudhuri P. Synthesis, CO_2 absorption property and densification of Li_4SiO_4 powder by glycine-nitrate solution combustion method and its comparison with solid state method. Journal of Alloys and Compounds. 2017;**725**:461

[8] Boudaren C. Dissertation, Constantine, Algeria. 2002

[9] Rao CNR, Nagarajan R, Vijayaraghaven R. Superconductor Science and Technology Synthesis of cuprate superconductors. Superconductor Science and Technology. 1993;**6**:1

[10] Rosales BA, Wei L, Vela J. Synthesis and mixing of complex halide perovskites by solvent-free solid-state methods. Journal of Solid State Chemistry. 2019;**271**:206

[11] Vlášková K, Colman RH, Klicpera M. Synthesis of $Er_2Ir_2O_7$ pyrochlore iridate by solid- state-reaction and CsCl flux method. Materials Chemistry and Physics. 2021;**258**:123868

[12] Tarasova N, Animitsa I, Denisova T, Nevmyvako R. The influence of fluorine doping on short-range structure in brownmillerite $Ba_{1.9}5In_2O_{4.9}F_{0.1}$. Solid State Ionics. 2015;**275**:47

[13] Patil KC, Aruna ST, Ekambaram S. Combustion synthesis. In Current Opinion in Solid State and Materials Science. 1997;**2(2)**:158

[14] Bahadur D, Rajkumar S, Kumar A. Influence of fuel ratios on auto combustion synthesis of barium ferrite nano particles. Journal of Chemical Sciences. 2006;**118**:15

[15] Anuradha TV, Ranganathan S, Mimani T, Patil KC. Combustion synthesis of nanostructured barium titanate. Scripta Materialia. 2001;**44**:2237

[16] Sharma A, Modi OP, Gupta GK. Combustion Synthesis of Nanocrystalline Al_2O_3 Powder using Aluminium Nitrate and Urea as reactants-influence of reactant composition. Advances in Applied Science Research. 2012;**3(6)**:3819

[17] Mukasyan AS, Rogachev AS. Discrete reaction waves: Gasless combustion of solid powder mixtures. Progress in Energy and Combustion Science. 2008;**34**:377

[18] Ford RW. Ceramics Drying. New York: Pergamon; 1986

[19] Brinker CG, Scherer GW. Sol-Gel Science: The Physics and Chemistry of Sol-Gel Processing. Boston: Academic Press; 1990

[20] Livage J. Sol-gel synthesis of solids. Encyclopedia of Inorganic Chemistry, New York: John Wiley; 1994

[21] Pierre AC. Sol Gel Technology. In Kirk-Othmer Encyclopedia of Chemical Technology. Weinheim: John Wiley; 2007

[22] Landau MV. Sol-Gel Processing. In Synthesis of Solid Catalysts. Weinheim: Wiley- VCH; 2009

[23] Kumar A, Yadav N, Bhatt M, Mishra NK et al. Sol-Gel Derived Nanomaterials and It's Applications: A Review. Research Journal of Chemical Sciences. 2015;**5(12)**:98

[24] Fletcher JM, Hardy CJ. Application of sol-gel processes to industrial oxides. Chemistry, and Industry. 1968;**2**:48

[25] Triki M, Ksibi Z, Ghorbel A, Medina F. Preparation and characterization of CeO_2-TiO_2 support for Ru catalysts: Application in CWAO of phydroxybenzoic Acid. Microporous and Mesoporous Materials. 2009;**117**:431

[26] Hammedi T, Triki M, Ksibi Z, Ghorbel A, Medina F. Comparative study of textural, structural and catalytic properties of xerogels and aerogels CeO_2-TiO_2 mixed oxides. Journal of Porous Mater. 2015; **22**:939

[27] Haruna A, Abdulkadir I, Idris SO. Synthesis, characterization and photocatalytic properties of $Bi_{0.85-X}MXBa_{0.15}FeO_3$ (M = Na and K, X = 0, 0.1) perovskite-like nanoparticles using the sol-gel method. Journal of King Saud University Science. 2020;**32(1)**:896

[28] Ahmoum H, Li G, Piao Y, Liu S et al. Ab-initio, Monte Carlo and experimental investigation on structural, electronic and magnetic properties of $Zn_{1-x}Ni_xO$ nanoparticles prepared via sol-gel method. Journal of Alloys and Compounds. 2021;**854**:157142

[29] Zhang Q, Lu H, Wang Y, Wu Y et al. Synthesis of $La_2Mo_2O_9$ Nanocrystallites by Sol-Gel Process. Asian Journal of Chemistry. 2014;**26**:1334

[30] Tazikeh S, Akbari A, Talebi A, Talebi E. Synthesis and characterization of tin oxide nanoparticles via the Co-precipitation method. Materials Science-Poland. 2014;**32(1)**:98

[31] Murray CB, Kagan CR, Bawendi MG. Synthesis and Characterization of Monodisperse Nanocrystals and Close-Packed Nanocrystal Assemblies. Annual Review of Materials Science. 2000;**30**:545

[32] Zhu X, Du Y, Wang H, Wei Y et al. Chemical interaction of Ce-Fe mixed oxides for methane selective oxidation. Journal of Rare Earths. 2014;**32**:824

[33] Polsongkram D. Chamninok P, Pukird S, Chow L et al. Effect of synthesis conditions on the growth of ZnO nanorods via hydrothermal method. Physica B: Condensed Matter. 2008;**403**:3713

[34] Liu HS, Chin TS, Lai LS, Chiu SY et al. Hydroxyapatite synthesized by a simplified hydrothermal method. Ceramics International. 1997;**23**:19

[35] Kothapalli CR. Wei M, Legeros RZ, Shaw MT. Influence of temperature and aging time on HA synthesized by the hydrothermal method. Journal of Materials Science: Materials in Medicine. 2005;**16**:441

[36] Li N, Yanagisawa K. Controlling the morphology of yttrium oxide through different precursors synthesized by hydrothermal method. Journal of Solid State Chemistry. 2008;**181**:1738

[37] Triki M, Kochkar H, Berhault G, Ghorbel A. Highly active ruthenium catalysts supported on nanostructured titanates for application in catalytic wet air oxidation of p- hydroxybenzoic acid. Reaction Kinetics, Mechanisms and Catalysis. 2010;**101**:377

[38] Jemmali M, Walha S, Ben Hassen R, Noel H. Synthesis and Crystal Structure of $Er_6Fe_{17.66}Al_{5.34}C_{0.65}$ Intermetallic Compound. X-ray Structure Analysis Online 2009;**25**:123

[39] Jemmali M, Walha S, Noel H, Ben Hassen R. Crystal structures and magnetic properties of the $RFe_{1-x}Ge_2$ (R = Gd, Er) compounds. Chemistry of Metals and Alloys. 2010;**3**:126

[40] Walha S, Jemmali M, Skini RH, Noel H et al. Structural and Magnetocaloric Properties of $ErFe_{2.4}Al_{0.6}$ Compound. Journal of Superconductivity and Novel Magnetism. 2014;**27**:2131

[41] Nouri K, Saidi M, Walha S, Bessais L, Jemmali M. Magnetocaloric Effect in $SmNi_2$ Compound Chemistry Africa. 2020;**3**:111

[42] Nouri K, Jemmali M, Walha S, Zehani K, Ben Salah A, Bessais L. Structural, atomic Hirschfeld surface, magnetic and magnetocaloric properties of $SmNi_5$ compound. Journal of Alloys and Compounds. 2016;**672**:440

[43] Saidi M, Nouri K, Walha S, Dhahri E, Kabadou A, Jemmali M, Bessais L. Structural, Magnetic, Magnetocaloric and Mo¨ssbauer Spectrometry Study of $Gd_2Fe_{17-x}Cu_x$ (x =0, 0.5, 1 and 1.5) Compounds. Journal of Electronic Materials 2019 ;**48** ; 2242.

[44] Bouchaala N, Jemmali M, Bartoli T, Kamel N, Hentech I, Walha S, Salah AB, BessaisL. Infuence of Fe-substitution on magnetic, magnetocaloric, and Mossbauer effect study of $Nd_2Fe_{17-x}Co_x$ solid solutions. Journal of Solid State Chemistry. 2018; **258**; 501.

[45] Jemmali M, Alleg S, Dhahri E, Bessais L. Effect of Al Substitution on Structural, Magnetic and Magnetocaloric Properties of $Er_6Fe_{23-x}Al_x$ (x = 0 and 3) Intermetallic Compounds. Crystals. 2017; 7; 156.

[46] kamel N, Bouzidi w, Jemmali M, Hentech I, Dhahri E, Bessais L. Effect of ball-milling on magnetic properties of uniaxial nanocrystalline SmNi2Fe Journal of Electronic Materials. 2017; **46**; 1971.

[47] Jemmali M, Nouriz K, Benali A, Walha S, Salah A. Ben, Dhahri E and Bessais L. Effect of small Fe content on the structure, magnetic and magnetocaloric properties of $SmNi_{3-x}Fe_x$ (x = 0; 0,3 and 0,8) intermetallic compounds. J Supercond Nov Magn, 2018 ;**31**; 511.

[48] Saidi M, Walha S, Hlil E.K, Bessais L, Jemmali M. Effect of chromium substitution on structural, magnetic and magnetocaloric properties of $GdFe_{12-x}Cr_x$ intermetallic compounds, Mössbauer spectrometry and ab initio calculations. Journal of Solid State Chemistry, 2021; **297**; 122019

[49] Saidi M, Walha S, Nouri K, Kabadou M. Jemmali, L. Bessais. The 1073 K isothermal section of the Gd-Fe-Cu system. Journal of Alloys and Compounds. 2019;**781**:159.

[50] Saidi M, Walha S, Nouri K, Kabadou A, Bessais L, Jemmalil M. An investigation of the Gd-Fe-Cr phase diagram: Phase equilibria at 800°C. Journal of Alloys and Compounds. 2019; **792**: 87.

[51] Nouri K, Jemmali M, Walha S, Salah AB. Dhahri E and Bessais L. Experimental investigation of the Y-Fe-Ga ternary phase diagram :Phase equilibria and new isothermal section at 800 °C. . Journal of Alloys and Copounds 2017;**719**:256

[52] Bouchaala N, Jemmali M, Nouri K, Walha S. Zehania K, Salah AB, Bessais L. Experimental Investigation of the Isothermal Section at 800 °C of the Nd-Fe-Co Ternary System Journal of Phase Equilibria and Diffusion. 2017;**38**:561

[53] Nouri K, Jemmali M, Walha S, Zehani K. Bessais L, Salah AB. The isothermal section phase diagram of the Sm-Fe-Ni ternary system at 800°C Journal of Alloys and Compounds. 2016;**661**:508

[54] Jemmali M, Walha S, Ben Hassen R, Noel H. Isothemal cross-sections of the Gd-Fe- Ge system at 800°C and Magnetic properties of Gd117Fe52Ge112 compound. Asian Journal of Chemistry. 2016;**28**:1330

[55] Jemmali M, Walha S, Pasturel M, Tougait O, Ben Hassen R, Noel H. Isothermal section of the Er-Fe-Al ternary system at 800°C . Journal of Alloys and Compounds. 2010; **489**: 421

[56] Saidi M, Nouri K, Walha S, Bessais L, Jemmali M. Solid-state phase equilibria in the Er-Nd-Fe ternary system at 1073 K. Journal of Alloys and Compounds. 2020;**844**:155754

[57] Cizmasa CB, Sab S, Bessais L, Djega-Mariadassou C. Magnetocrystalline anisotropy in $PrFe_{11-x}Co_xTiC$. Journal of Magnetism and Magnetic Materials. 2004;**272**:e395

[58] Cizmas CB, Bessais L, Djega-Mariadassou C, Meyer C, Voiron J. Effect of carbonation on the structure and magnetic properties of $YFe_{11-x}Co_xTiC_y$ alloys. Journal of Magnetism and Magnetic Materials. 2007;**310(2)**:e623

[59] Molina-Boisseau S, Le Bolay N. Size reduction of polystyrene in a shaker bead mill- kinetic aspects. Chemical Engineering Journal. 2000;**79**:31

[60] Pan J, Shaw WJD. Effects of processing parameters on material properties of mechanically processed polyamide. Journal of Applied Polymer Science. 1995;**56**:557

[61] Suryanarayana C. Mechanical alloying and milling. Mechanical alloying and milling, Progress Mater. Sci. 2001;**46**:1

[62] Phejar M, Paul-Boncour V, Bessais L. Structural and magnetic properties of magnetocaloric $LaFe_{13-x}$ Si_x compounds synthesized by high energy ball-milling. Intermetallics. 2010;**18**:2301

Chapter 2

Effect of Operating Parameters and Foreign Ions on the Crystal Growth of Calcium Carbonate during Scale Formation: An Overview

Atef Korchef

Abstract

Due to the insufficiency of freshwater resources and to supply the drinking water populations, many desalination processes such as reverse osmosis, electrodialysis and distillation are widely used. However, these processes are of large-scale consumers of energy and confronted with various problems such as corrosion and scale formation. In most cases, scales are made of calcium carbonate $CaCO_3$. In the present chapter, an overview on the effect of operating parameters such as temperature, pH and supersaturation on the precipitation kinetics, microstructure, and polymorphism of $CaCO_3$ is given. Additionally, I put special emphasis on the effect of foreign ions such as magnesium, sulphate, and iron ions on $CaCO_3$ precipitation since they are present at significant concentrations in natural waters. Also, the mechanisms by which these ions affect the crystal growth of $CaCO_3$ were pointed out. Knowledge about these operating parameters as well as the effects of foreign ions allow elucidating the polymorphs growth during water treatment. The control of these operating parameters allows reducing scale formation during drinking water and wastewater treatment. The economic impact is of greatest importance since this favorably affects the treatment costs, increases the equipment life, and allows enhanced product water recovery.

Keywords: water treatment, scale, calcium carbonate, cement, growth, magnesium, sulphate, iron, CO_2, supersaturation, pH, temperature

1. Introduction

Calcium carbonate $CaCO_3$ is a commercial material and a major constituent in many natural systems. $CaCO_3$ is widespread in the natural environment such as eggshells, corals, and sedimentary rocks [1–3]. $CaCO_3$ is applied as filler and pigment in the production of paper, rubber, plastic, paint, textiles, and pharmaceuticals [4, 5]. More interesting, $CaCO_3$ is used in ionic cements which are widely used in bone and dental replacement biomaterials. Original compositions of bioresorbable and biocompatible cements including a significant proportion of synthetic $CaCO_3$ have been developed to meet a need for biomedical cements with increased resorption properties. Bone substitutes based on chemically treated natural $CaCO_3$ are biocompatible and bioactive. They are used in the form of powder and porous ceramics [6–11]. Because of the higher solubility of calcium carbonates compared to calcium phosphates, the introduction of a significant amount of a metastable variety of $CaCO_3$ (vaterite or aragonite) should give mineral cements a higher rate of resorption and thus promote faster bone reconstruction [11, 12].

On the other hand, $CaCO_3$ has a significant impact on energy production and water treatment. The insufficiency of the freshwater resources and the requirements of drinking water will be increasingly manifest in the years to come. It is very probable that the problem of water, just like that of the energy resources, will be regarded as one of the determining factors of the stability of a country. The surest and economic means to supply the drinking water populations require the desalination of brackish, saline or sea waters by using different processes of desalination (reverse osmosis, electrodialysis, distillation, etc). However, these processes are generally of large-scale consumers of energy and confronted with various problems, such as corrosion and scale formation (**Figure 1**) which cause enormous energy losses. In most cases, scales are made of the sparingly soluble salt $CaCO_3$ [13–19]. Because of its poor thermal conductivity and its good adherence to the walls, $CaCO_3$ decreases the heat transfer rate, reduces the water flow rate, and even shortens equipment life by corrosion [20, 21]. The crystallization of $CaCO_3$ depends on several operating parameters such as the mineralogical water composition, the supersaturation of the treated water, the pH, and the temperature.

In the present overview chapter, I was interested in the scale problem through $CaCO_3$ precipitation encountered during drinking water and wastewater treatment plants. The effect of the operating parameters such as temperature, pH, supersaturation, and foreign ions on the $CaCO_3$ crystal growth, microstructure and polymorphism was exposed. Knowledge about these operating parameters for $CaCO_3$ crystallization as well as the effects of foreign ions, especially magnesium, sulphate, and iron ions, is very important in the elucidation of the $CaCO_3$ polymorphs growth during water treatment. The presence of mineral ions can have a major influence on the crystal growth and microstructure of $CaCO_3$ since they are present with significant concentrations in natural waters. The economic impact is of the utmost importance. Indeed, the control of the operating parameters makes it possible to inhibit or reduce the scale formation during the treatment of drinking water and wastewater in different industrial processes such as desalination units, water pumps

Figure 1.
Scaling through $CaCO_3$ precipitation in drinking water pipes.

and heat exchangers. This positively affects processing costs, increases the life of the equipment, and enhances the product water recovery.

2. Polymorphism of calcium carbonate

Calcium carbonate has six different polymorphs: one amorphous $CaCO_3$, two hydrated crystalline ones (hexahydrate $CaCO_3{\cdot}6H_2O$ and monohydrate $CaCO_3{\cdot}H_2O$), and three anhydrous crystalline polymorphs that are the hexagonal vaterite, the orthorhombic aragonite and the rhombohedra calcite [22–25]. The X-ray diffraction (XRD) diagrams of these polymorphs are shown in **Figure 2**. For $CaCO_3$ polymorphs, the solubility increases in the order of calcite, aragonite, and vaterite [26, 27]. Calcite is the most stable polymorph under ambient atmospheric conditions [26]. In the presence of water, vaterite, which is the least thermodynamically stable polymorph, easily transforms into the more stable calcite or aragonite [28]. Environmental conditions play effective roles in crystal growth and morphological changes, namely, the solution supersaturation, calcium concentration, pH, and temperature. These experimental parameters should operate together and control the polymorphism of $CaCO_3$.

3. System $CaCO_3$-H_2O-CO_2

Scaling is a complex phenomenon that takes a long time to manifest itself in industrial or domestic facilities. Several techniques, such as mechanical stirring [29], aeration [30], electrochemical precipitation [31], magnetic water treatment [32] have been used in laboratories to assess the effectiveness of chemical or physical treatment, and the influence of certain parameters on scale formation over time. However, the most interesting method used for $CaCO_3$ precipitation is the CO_2 repelling method [16, 17, 33] where the CO_2 containing solution is air bubbled which increases the solution pH and provokes the $CaCO_3$ precipitation (**Figure 3**). Indeed, this method is an accelerated simulation of the natural scaling phenomenon

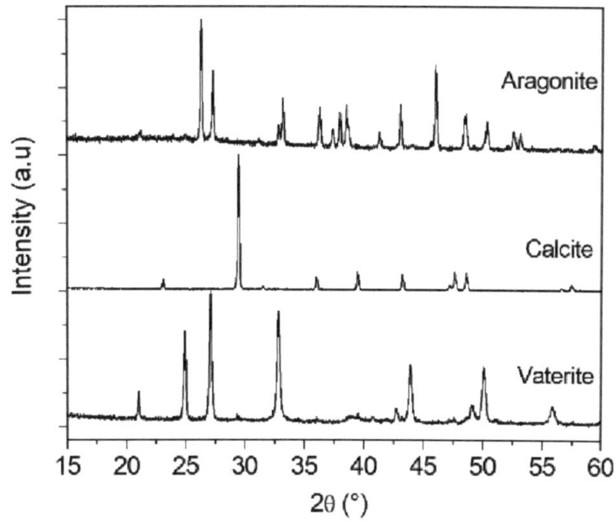

Figure 2.
XRD diagrams of vaterite, calcite and aragonite.

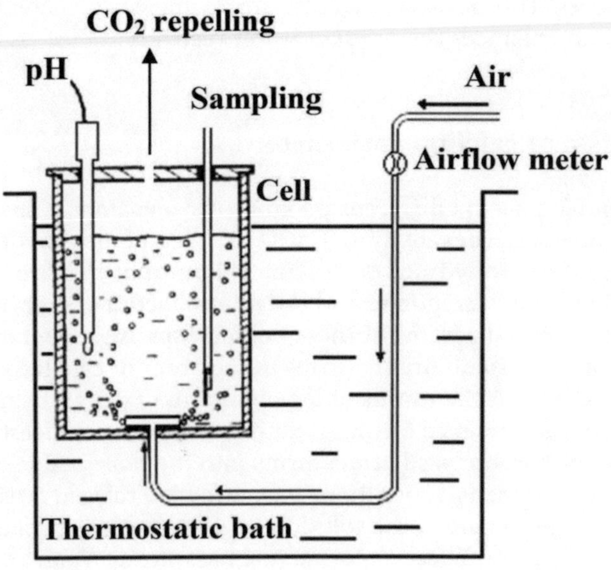

Figure 3.
Schematic illustration of the CO_2 repelling method used for $CaCO_3$ precipitation. A cylindrical cell containing a diffuser at the bottom and covered with a lid. The lid contains different openings, one of which allows samples to be removed to determine the calcium amount left in the solution and another opening allows CO_2 repelling from the solution while the air is bubbling. The flow of the injected air was controlled by an air flow meter, and the temperature of the solution was controlled by a thermostat with circulating water. An electrode is used to record the pH.

in which the precipitation of $CaCO_3$ takes place, in a short experiment time that do not exceed 90 min, following the removal of the dissolved CO_2 by the atmospheric air, according to the following reaction [16, 17]:

$$Ca^{2+} + 2HCO_3^- \rightarrow CaCO_3 + CO_2 + H_2O \tag{1}$$

The saturation index (SI) is a measure of the deviation of the system from equilibrium. When SI < 0, the solution is undersaturated and no crystallization occurs. When SI = 0, the solution is in equilibrium, and when SI > 0, the solution is supersaturated, and crystallization could occur spontaneously. For $CaCO_3$, SI is given by the following equation:

$$SI = log\,\Omega \tag{2}$$

where $\Omega = \frac{IAP}{K_{sp}} = \frac{(Ca^{2+})(CO_3^{2-})}{K_{sp}}$ is the supersaturation ratio of the solution with respect to $CaCO_3$ and IAP and K_{sp} represent respectively, the ionic activity product and the thermodynamic solubility product of $CaCO_3$. The solubility products of the different polymorphs of $CaCO_3$, given in **Table 1**, were determined by the temperature dependent equations given by Plummer and Busenberg [27] and Brecevic and Kralj [4].$(Ca^{2+}) = \gamma_{Ca^{2+}}[Ca^{2+}]$ and $(CO_3^{2-}) = \gamma_{CO_3^{2-}}[CO_3^{2-}]$ are the activities of the ions Ca^{2+} and $CO_3{}^{2-}$, respectively. The activity coefficients $\gamma_{Ca^{2+}}$ and $\gamma_{CO_3^{2-}}$ are calculated using the extended Debye-Huckel equation:

$$log\,\gamma_i = -\frac{AZ_i^2\sqrt{I}}{1 + Ba_i\sqrt{I}} + b_i\sqrt{I} \tag{3}$$

Polymorphs	−log Ks	T in the range	Ref.
Calcite	171.9065 + 0.077993 T − 2839.319/T − 71.595 log(T), T in K	0–90°C	[27]
Vaterite	172.1295 + 0.077996 T − 3074.688/T − 71.595 log(T), T in K	0–90°C	[27]
Aragonite	171.9773 + 0.077993 T − 2903.293/T − 71.595 log(T), T in K	0–90°C	[27]
Amorphous	6.1987 + 0.0053369 T + 0.000109 T^2, T in °C	10–55°C	[4]
$CaCO_3{\cdot}H_2O$	7.050 + 0.000159 T^2, T in °C	10–50°C	[4]
$CaCO_3{\cdot}6H_2O$	7.1199 + 0.011756 T + 0.000075556 T^2, T in °C	10–25°C	[4]

Table 1.
Logarithmic solubility products of the various polymorphs of calcium carbonate $CaCO_3$ as a function of temperature.

where A, B are constants defined by Harmer [34], a_i and b_i are ion-specific parameters assigned by Truesdell and Jones [35], Z_i is the valence of the i-species and I is the ionic strength:

$$I = \frac{1}{2}\sum_i Z_i^2 c_i \tag{4}$$

where c_i is the concentration of the i-species.

In the carbonation process, the CO_2 gas firstly dissolves into water to form the carbonic acid H_2CO_3 which transforms into HCO_3^-, CO_3^{2-} and H^+ according to the following equations:

$$CO_2(g) \leftrightarrow CO_2(lq), \quad P_{CO_2} = K_H(CO_2) \tag{5}$$

$$CO_2(lq) + H_2O \leftrightarrow H_2CO_3 \tag{6}$$

$$H_2CO_3 \leftrightarrow HCO_3^- + H^+, \quad \frac{(HCO_3^-)(H^+)}{(CO_2)} = 10^{-K_1} \tag{7}$$

$$HCO_3^- \leftrightarrow CO_3^{2-} + H^+, \quad \frac{(CO_3^{2-})(H^+)}{(HCO_3^-)} = 10^{-K_2} \tag{8}$$

where K_H, K_1 and K_2 are the Henry's law coefficient, first and second dissociation constants of carbonic acid (given in **Table 2** [36]), respectively and (HCO_3^-), (H^+) and (CO_2) are the activities of the ions HCO_3^-, H^+ and CO_2, respectively.

The electric neutrality of the calco-carbonic solution gives:

$$2[Ca^{2+}] + [H^+] = [OH^-] + [HCO_3^-] + 2[CO_3^{2-}] \tag{9}$$

and

$$(CO_3^{2-}) = \gamma_{CO_3^{2-}}[CO_3^{2-}] = \frac{1}{2}\gamma_{CO_3^{2-}}(2[Ca^{2+}] + [H^+] - [OH^-] - [HCO_3^-]) \tag{10}$$

From Eq. (8), the activity of CO_3^{2-} is determined by:

$$(CO_3^{2-}) = (HCO_3^-)10^{pH-K_2} = \gamma_{HCO_3^-}[HCO_3^-]10^{pH-K_2} \tag{11}$$

where $\gamma_{CO_3^{2-}}$ and $\gamma_{HCO_3^-}$ are the activities coefficients of the ions CO_3^{2-} and HCO_3^-, respectively. It is worth to note that the concentration of HCO_3^- ions can be assumed to be equal to the total alkali concentration (TAC) for pH in the range

Carbonation process		CO_2 (g) ↔ CO_2 (lq)	CO_2 (lq) + H_2O ↔ H_2CO_3 H_2CO_3 ↔ HCO_3^- + H^+	HCO_3^- ↔ CO_3^{2-} + H^+
Equilibrium equation		$P_{CO_2} = K_H(CO_2)$	$\frac{(HCO_3^-)(H^+)}{(CO_2)} = 10^{-K_1}$	$\frac{(CO_3^{2-})(H^+)}{(HCO_3^-)} = 10^{-K_2}$
Equilibrium constants		$K_H \times 10^{-5}$ ($PaLmol^{-1}$)	K_1	K_2
Temperature (°C)	25	22.549	6.4	10.35
	30	33.769	6.37	10.31
	40	42.586	6.34	10.24
	50	51.920	6.32	10.19
	60	61.476	6.33	10.16

Table 2.
Henry's law coefficient (K_H), and first and second dissociation constants (K_1, K_2) of carbonic acid for different temperatures (values are taken from Ref. [36]).

6–8.8. Indeed, $[CO_3^{2-}]$ and $[OH^-]$ are small in this pH range [37] and in a first approximation:

$$TAC = [HCO_3^-] + 2[CO_3^{2-}] + [OH^-] \approx [HCO_3^-] \tag{12}$$

4. Effect of operating parameters on $CaCO_3$ precipitation

The solubility of $CaCO_3$ polymorphs increases in the order of calcite, aragonite, and vaterite [26, 27]. Environmental conditions play effective roles in the crystal growth and morphological changes of $CaCO_3$. Indeed, the nucleation and growth kinetics of $CaCO_3$ as well as the morphology and polymorphism of the obtained precipitates are affected by various key operating parameters such as supersaturation [23, 28, 38], temperature [37, 39, 40], and pH [41–43]. These experimental parameters operate together and control the polymorphism of $CaCO_3$. For example, it was shown that vaterite can be obtained at high superasaturations and low calcium concentrations typically below 10^{-2} M [44], solution pH in the range 8.5–10 [45] and/or low temperatures in the range 20–40°C [45, 46].

Söhnel and Mullin [38] showed that the crystal growth rate of $CaCO_3$ increased with increasing the solution supersaturation. Vaterite was favored in moderate ($S < 6.5$) to enhanced supersaturation solutions [28, 44] and smaller vaterite crystals were obtained when supersaturation increased [28]. High supersaturations in the range 4−14 prevented the transformation of vaterite into calcite and increased the crystal growth rate [16, 23].

It is worth noticing that the initial calcium concentration in the solution has an important impact on the polymorphism of $CaCO_3$. Indeed, it was found that calcite did not crystallize at decreased Ca^{2+} concentrations (typically below 10^{-2} M), and only vaterite was obtained when $CaCO_3$ was prepared, at 25°C, by a CO_2/N_2 mixed gas bubbled into a solution of $CaCO_3$ [23]. This agrees with the results of Korchef [16] who showed that vaterite was the predominant polymorph at 28°C, for a supersaturation in the range 4–14, and calcite was not detected, because of the low initial calcium concentration (4×10^{-3} M). Thus, for a solution supersaturated with respect to calcite (~15–62), aragonite (4–17) and vaterite (10–38), vaterite was favored for an initial calcium concentration 4×10^{-3} M ($<10^{-2}$ M), a precipitation pH in the range 8.4–8.8 and a temperature of 28°C [16].

The temperature and the solution pH are considered as critical parameters in controlling the kinetics crystallization, polymorphism, and morphology of $CaCO_3$ particles. At room temperature, calcite is the most predominant phase. At temperatures higher than 50°C, $CaCO_3$ precipitates as needles-like aragonite crystals [47–50]. At conditions of spontaneous $CaCO_3$ formation preceded by the lapse of induction time, the increase of temperature, at fixed ionic strength, results to the decrease of the induction time and to the increase of the crystalline growth rate and also the amount precipitated [37]. Chen and Xiang [46] investigated the effect of temperature on the structures and morphology of $CaCO_3$ crystals obtained by double injection of $CaCl_2$ and NH_4HCO_3 solutions with a 1:1 molar ratio. At 30−40°C, they obtained crystals of lamellar vaterite. At higher temperatures of 50–70°C, a mixture of calcite and aragonite formed, and only aragonite whiskers crystallized at a temperature of 80°C. However, aragonite was not detected when $CaCO_3$ precipitated by mixing calcium acetate $Ca(C_2H_3O_2)_2$ and ammonium carbonate $(NH_4)_2CO_3$ [40].

In their recent publication, Korchef and Touabi [17] studied the effect of temperature and initial solution pH on the crystallization of $CaCO_3$ by CO_2 repelling. They showed that vaterite and aragonite were the polymorphs of $CaCO_3$ collected at 28°C and 50°C and the increase in temperature from 28–50°C leads to the acceleration of $CaCO_3$ nucleation and crystal growth. This is was explained by higher supersaturations with respect to aragonite (in the range ~ 40–69) and vaterite (between ~12 and 21) obtained at 50°C than those obtained at 28°C. The increase of the initial solution pH accelerates the $CaCO_3$ precipitation. Indeed, the solution supersaturation increases with increasing the initial pH of the solution. This accelerates the nucleation and growth of $CaCO_3$. For example, at the initial solution pH 7, the supersaturation ratios with respect to calcite and vaterite are in the ranges ~17–30 and ~5–8, respectively. For the initial solution pH 9, $CaCO_3$ precipitation is instantaneous and the supersaturation ratios with respect to calcite and vaterite significantly increase, i.e., in the range ~51–84 for calcite and in the range ~14 –23 for vaterite. However, at constant supersaturation, it was shown that the growth rate of calcite decreases with increasing pH from 7.5 to 12 [41]. This is because the surface concentration of active growth sites decreases with increasing pH of the solution. Ramakrishna et al. [42] prepared $CaCO_3$ crystals by mixing Na_2, CO_3, and $CaCl_2$ solutions injected simultaneously into distilled water at different pH values. They showed that pure aragonite needles were formed at pH 10 and a mixture of calcite and aragonite was obtained when increasing the pH from 11 to 12. At high pH values, typically greater than 12, calcite was favored [43], and at pH lower than 8, vaterite was obtained [23]. Using a constant composition method, Tai and Chen [45] showed that vaterite was the predominant phase obtained, at 24°C, in the pH range 8.5−10, which is comparable to the precipitation pH range obtained when $CaCO_3$ was precipitated by the CO_2 repelling method (8.4−8.8) [16, 17]. For solution pH lower than 7–7.5, no precipitation of $CaCO_3$ was detected [17]. This result is of utmost importance in inhibiting $CaCO_3$ precipitation during scale formation in water and wastewater treatment processes, as will be discussed in the last paragraph of the present chapter.

5. Effect of foreign ions on $CaCO_3$ precipitation

The presence of mineral ions can significantly influence the crystal growth and microstructure of $CaCO_3$ [50–59]. Among mineral ions, sulphate and magnesium ions are widely studied [50, 53–59] since they are present at significant concentrations in natural waters. Compared to sulphate and magnesium ions, the impact of iron on the precipitation of calcium carbonate seems to be less studied and it appears to be a lack of reliable data on its impact on $CaCO_3$ precipitation, especially at elevated temperature and different pH. **Figure 4** shows a tank highly containing iron-rich water. Iron ions can be present in the aqueous phase either from the corrosion of metallic parts in water desalination processes, such as tanks and pipes [60] or they may be present in groundwater at concentration levels depending on the mineral composition of the aquifer [61]. The water treatment processes are selected according to the water salt concentration. Indeed, membrane technological processes are used to desalinate brackish water containing salt concentrations below 10 g/L, however, reverse osmosis and thermal distillation are generally used to desalinate sea water with salt concentration above 30 g/L. Note that the typical iron concentration in the water used for the above mentioned processes is approximately 2.8 mg/L [62]. Groundwaters containing more than 0.2 mg/L of iron should be treated if used for drinking water [63]. Therefore, it is of great importance to understand how iron and calcium ions interact with each other during $CaCO_3$ precipitation. Throughout the followings, I put a special emphasis on the effects of magnesium, sulphate and iron ions on the precipitation kinetics, microstructure, and polymorphism of $CaCO_3$. Also, the mechanisms by which these ions affect the crystal growth of calcium carbonate were pointed out. Knowledge about the effects of these ions is very important in the elucidation of the growth polymorphs during water treatment.

5.1 Effect of magnesium and sulphate ions

It was shown that both sulphate and magnesium ions inhibit $CaCO_3$ crystallization [50, 55, 56], even though some discrepancies on the effect of sulphate ions were revealed. For example, at fixed temperature and ionic strength, Karoui et al. [50] showed that, the addition of sulphate ions to the $CaCO_3$ solution increased the induction time and decreased both the crystal growth rate and the amount of the precipitated $CaCO_3$. Comparable results were found by Vavouraki et al. [57]. However, Tlili et al. [64] showed that sulphate ions increase the growth rate of $CaCO_3$ precipitated electrochemically. Many works [51, 65, 66] have shown that sulphate ions inhibit vaterite transformation and enhance the calcite formation for low concentrations. At high concentrations, it was shown that sulphate ions

Figure 4.
A tank containing iron-rich water. Iron ions can be present in the aqueous phase either from the corrosion of metallic parts or they may be present in groundwater at concentration levels depending on the mineral composition of the aquifer.

(i) increased the nucleation time [50], (ii) decreased the precipitation rate of calcite formation [57] and (iii) promoted the formation of aragonite via a dissolution–precipitation process after long reaction times [55, 59].

The temperature and the ratio of magnesium to calcium control the precipitation kinetics, the type of polymorph and the morphology of the $CaCO_3$ precipitates [37, 64, 67]. For example, it was shown that when the Mg^{2+}/Ca^{2+} ratio passes from 0 to 4, with a Ca^{2+}concentration of 4×10^{-3} mol L^{-1}, the scaling time increases from 50 to 400 min, respectively [64]. Mejri et al. [37] studied the effect of magnesium ions on the $CaCO_3$ precipitation, by the CO_2 repelling method, at different temperatures between 30°C and 60°C and with different magnesium to calcium molar ratios (R) in the range 2–5. They showed that, at a fixed temperature, the increase of Mg^{2+} concentration significantly increased the induction time (t_n) and decreased both the initial crystal growth rate (V_i) and the amount precipitated (Rp). For example, at 40°C, t_n passed from 7 to 30 min, V_i passed from 0.57 to 0.49 M min^{-1} and Rp decreased from 81 to 56% when R passed from 0 to 5, respectively. The increase in temperature from 30–60°C weakens the impact of magnesium ions on the retardation of $CaCO_3$ precipitation. These results agree with previous works [38, 67–69], showing that the induction time significantly increased due to the presence of Mg^{2+}. At high concentrations of magnesium, amorphous $CaCO_3$ exhibited a prolonged stability, while it transformed instantly to calcite and vaterite in pure water [58]. In the presence of high amounts of Mg^{2+}, $CaCO_3$ exclusively precipitated as aragonite while calcite and vaterite formed when no Mg^{2+} ions are present [70]. At low temperature of 30°C, the addition of Mg^{2+} ions favored the crystallization in the bulk solution of secondary aragonite, resulted from the transformation of vaterite nuclei, while primary aragonite was favored, at high temperature of 60°C, with larger particles sizes than those obtained at 30°C [37].

5.2 Effect of iron ions

The results reported in literature on the effect of iron ions on $CaCO_3$ precipitation are often contradictory. For example, Kelland [71] found that iron ions concentrations below 25 ppm did not affect the $CaCO_3$ precipitation in high pressure dynamic tube blocking tests. Comparable results were found by Lorenzo et al. [72] who studied the effect of ferrous iron ions on the $CaCO_3$ growth by the constant composition method. However, Herzog et al. [73] found that, in magnetic water treatment devices, an excess of ferrous ions (5.6 ppm) strongly inhibits both calcite growth and the transformation of aragonite to calcite. The inhibiting effectiveness depends strongly on the solution supersaturation [16, 17, 74–76]. Indeed, at high supersaturations, the inhibition effectiveness of iron on $CaCO_3$ crystallization is small. In that case, the inhibition effectiveness can be improved by lowering the solution pH, increasing the iron concentration and/or by lowering the solution supersaturation [16, 17]. At relatively low supersaturations, the addition of iron ions retards the $CaCO_3$ precipitation. Past the onset of nucleation, iron ions enhance the growth rate of $CaCO_3$ and most of $CaCO_3$ amounts precipitate in the bulk solution instead of on the cell walls minimizing, therefore, the risks of scaling. Scaling is defined as the precipitation on the cell walls [16]. These results agree with those of Katz et al. [74] and Takasaki et al. [76] showing that lower supersaturations require less iron for the same extent of $CaCO_3$ growth inhibition.

It was shown [16] that for both temperatures 28°C and 50°C, the addition of iron retarded the nucleation of $CaCO_3$ and, most important, it decreased the $CaCO_3$ precipitation in the cell walls which can be easily collected, reducing therefore scale phenomena in water treatment devices. At a fixed temperature of 50°C, the increase in the iron concentration in the solution to 0.5 mg/FeL did not significantly affect the $CaCO_3$ precipitation [17]. Comparable result was obtained by Macadam and Parsons [77] who showed that addition of 0.5 mg L^{-1} of iron did not affect the induction time of $CaCO_3$ precipitation by magnetic stirring at 40°C. For higher iron concentrations, the solution supersaturation decreased, and the inhibition effectiveness of iron ions became more pronounced [17]. At 50°C, the addition of iron ions in the solution promoted aragonite formation rather than vaterite. Indeed, with the increase in the iron ions concentration, the solution supersaturation decreased and the dissolution of vaterite became, most likely, faster which promotes aragonite formation. However, at 28°C, the addition of iron ions did not affect the nature of the precipitated phases vaterite and aragonite, and vaterite with rough surfaces was the predominant polymorph. This was explained that iron ions substituted partially Ca^{2+}, which inhibits vaterite dissolution and prevents the transformation of vaterite into aragonite [16, 17].

For a given iron concentration, increasing the initial pH causes the decrease in the induction time and the increase in the initial crystal growth rate. However, increasing iron ions concentration at a fixed pH in the range 7–9, results in the decrease in both the induction time and the crystal growth rate of $CaCO_3$ [17]. Additionally, the increase in iron ions concentration at a fixed pH reduced the amount of the precipitates obtained and, however, enhanced the precipitation of $CaCO_3$ in the bulk solution, reducing therefore the risks of scaling. Therefore, the $CaCO_3$ precipitation on cell walls, which causes the major phenomenon of scaling, could be lowered by enhancing the initial solution pH or by controlling the concentration of iron ions in the solution. However, the increase of the initial solution pH leads to high amounts of $CaCO_3$ precipitates and even when formed in the bulk solution, they can agglomerate and block up different parts of the water treatment devices such as pipes and conducts, resulting in the blockage of fluid cooling. This increases processing costs, decreases equipment life, and decreases the product water recovery.

At 28°C and initial pH 8, calcite and vaterite were obtained. The increase in iron concentration in the solution resulted in an increased formation of calcite. At initial pH 9 and iron concentration of 4 mg/FeL, calcite was the only phase detected. At 28°C, the increase of the initial solution pH resulted in the disappearance of aragonite and the precipitation of vaterite and calcite. Iron ions addition up to 4 mg/FeL accelerated the transformation of vaterite into calcite and only calcite with stepped surfaces (**Figure 5**) was obtained at initial pH 9 [17]. The stepped surfaces were explained by iron ions adsorption or to difficulties in incorporating building units into the surfaces of calcite crystals [78].

5.3 Effect of chloride ions

For high concentrations, chloride might affect the stability of calcium in the solution, and thus the precipitation of $CaCO_3$. However, at relatively low temperatures between 15°C and 85°C, the equilibrium association constants of calcium ions with chloride ions in aqueous solution reported in literature are small. In fact, $\log K$ varies between −0.28 and −0.04 [79–83] which indicates, therefore, the instability of the ion pairs. Korchef and Touabi [17] showed that, at 28 and 50°C and for low concentrations of chloride below 0.18 mM, calcium chloride ion pairs either do not form or have stability fields too small to remarkably decrease the free calcium ion concentration in the solution and to reduce, therefore, the $CaCO_3$ growth rate. Calcium chloride ion pairs form at higher temperatures, i.e., for temperatures in the range 100-360°C, the stability field for $CaCl^+$ decreases with enhancing temperature, whereas that for $CaCl_2^0$ increases significantly [84]. It should be noted that higher order calcium chloride ion pairs do not form or have small stability fields.

Figure 5.
Calcite with stepped surfaces obtained at 28°C, initial solution pH 9, and iron ions concentration of 4 mg/FeL.

In addition, chloride ions can be added as sodium chloride NaCl. Tai and Chen [45] studied the effect of ionic strength and additives concentration on the polymorphism of $CaCO_3$ formed in a constant composition environment. They adjusted the ionic force by NaCl addition which is, they found, inactive in the polymorphic form. Moreover, Takita et al. [85] showed that low concentrations of NaCl in the range 0.1–0.5 M did not remarkably affect the solubility of $CaCO_3$. However, when an excess of NaCl (2.5 M) was added, the solubility of $CaCO_3$ increased which promoted the vaterite to calcite transformation. The resulting high concentration of $CaCO_3$ favored crystal growth rather than nucleation.

5.4 Effect of additives and dissolved organic matter

Ukrainczyk et al. [86] investigated the effect of salicylic acid derivatives on calcite precipitation kinetics and morphology. Their results showed that the adsorption of additive molecules lowered the $CaCO_3$ growth rate by blocking the propagation of growth sites. This causes the formation of steps and jagged and discontinuous surfaces. Comparable results were obtained for magnesium-containing solutions, where calcite crystal edges were rough and growth steps were apparent [55]. Comparable effects were also found when iron was added [16, 17]. The presence of additives, such as monoethylene glycol (MEG), prevents the vaterite to calcite transformation and stabilizes the formation of the relatively unstable vaterite polymorph. For example, Natsi et al. [87] showed that, in the presence of low concentrations of MEG of (10−20% v/v), vaterite was stabilized. At stable supersaturated solutions, seeded with quartz and calcite crystals, the growth of $CaCO_3$ was significantly reduced with the increase of MEG concentrations. At 60% of MEG, the vaterite/aragonite- to-calcite transformation was prevented, whatever the synthesis temperature [88].

On the other hand, it was shown that dissolved organic matter (DOM) inhibits $CaCO_3$ crystal growth [89]. The inhibition effect was related to the formation of calcium complexes and the occupation of the $CaCO_3$ growth sites which resulted in $CaCO_3$ rough surfaces. Dzacula et al. [90] studied in vitro precipitation of aragonite in artificial seawater at a high supersaturation of 11, and at a low supersaturation of 5.8. In either chemical systems, different concentrations of soluble organic matter (SOM), extracted from the symbiotic coral B. europea (SOM-Beu) and the a-symbiotic one L. pruvoti (SOMLpr) were added in order to investigate their effect on crystal growth or nucleation processes. They showed that, at high supersaturation, the SOMs increased the induction time but did not change the growth rate, and they were incorporated within nanoparticles aggregates. At low supersaturation, the SOMs affected the overgrowing crystalline unit aggregation and did not substantially affect the growth rate.

6. Possible mechanisms

According to the Ostwald [91] law, the least stable phase, that has the highest solubility, precipitates first and subsequently transforms to the more stable one. This transformation can be either a solid-state transition that invokes internal rearrangements of atoms, ions or molecules [92, 93] or a solution mediated transformation that invokes the dissolution of the metastable phase in the solution and simultaneous precipitation of the stable phase [94, 95]. It was shown that the amorphous calcium carbonate is a precursor in spontaneous precipitation of $CaCO_3$ at relatively high supersaturations. The amorphous phase, which consists of spherical particles with the diameter in the range 50–400 nm, dissolves rapidly in an undersaturated solution and it undergoes a rapid transformation to one of the more stable anhydrous forms (calcite, vaterite and aragonite) [4]. Therefore, the unstable amorphous calcium carbonate is firstly formed [36]. Then, it transforms into crystalline phases which are vaterite, calcite and aragonite. The vaterite can be transformed gradually into calcite or aragonite following two steps that are (i) the vaterite dissolution and (ii) the calcite or aragonite growth [71]. The second step controlled the overall rate of vaterite transformation to the most stable phase calcite or aragonite [96]. The two hydrated crystalline forms of calcium carbonate, calcium carbonate hexahydrate $CaCO_3{\cdot}6H_2O$ and calcium carbonate monohydrate $CaCO_3{\cdot}H_2O$, are more stable than the amorphous $CaCO_3$. Calcium carbonate monohydrate crystallized in well-defined spherical crystals with diameters in the range 15–30 mm and calcium carbonate hexahydrate crystallizes in well-defined rhombohedral crystals in the size range between 10 and 40 mm. Generally, hydrated forms precipitate from supersaturated solutions before the more stable anhydrous forms calcite, vaterite and aragonite [4].

As shown above, both sulphate and magnesium ions inhibit $CaCO_3$ crystallization [50, 55, 56]. The inhibition effect of magnesium ions was explained by (i) the increase in calcite solubility induced by magnesium ions insertion into the calcite crystals by substituting calcium ions [97]. This substitution depends on the Mg^{2+}, Ca^{2+} and SO_4^{2-} ions concentrations [50], (ii) accumulated strain provoked by smaller magnesium ions incorporation [50, 56] and (iii) adsorption of magnesium ions on calcite surfaces [56]. Indeed, based on the calculation of defect-forming energies in the crystalline lattice of aragonite, Mandakis et al. [98] showed that Mg^{2+} ions can substitute Ca^{2+} ions. This substitution may take place in the volume of the orthorhombic lattice of aragonite or on the crystal surfaces. Assessing the mechanism by which Mg^{2+} ions act to inhibit the crystalline growth of $CaCO_3$ requires an understanding of the changes that must occur when an ion is transferred from the solution to its location in the crystalline lattice. In the solution, Mg^{2+} ions are surrounded by six molecules of H_2O [99]. The incorporation of Mg^{2+} ions implies therefore the replacement of the H_2O molecules by six carbonate groups of the $CaCO_3$ lattice. This is done by partial dehydration (replacement of three H_2O molecules by oxygen from the calcite lattice) and then diffusion to a growth site and total dehydration (replacement of the three remaining H_2O molecules) [99]. According to Folk [100], the adsorption and incorporation of Mg^{2+} ions into the calcite lattice leads to the distortion of the crystalline lattice because of the lower ion radius of Mg^{2+} than that of Ca^{2+} (**Figure 6**). Thus distorted, the crystalline structure of the calcite can no longer accept the following Ca^{2+}, which comes to seek its position in the $CaCO_3$ lattice. Therefore, the crystalline growth of calcite is inhibited. This results in a solubility of calcite containing Mg^{2+} ions greater than that of pure calcite (without Mg^{2+}) [101].

Sulphate ions can be incorporated into the $CaCO_3$ lattice structure. Kitano et al. [97] demonstrated that this incorporation is easier in calcite than in aragonite and that with the increase in NaCl concentration in the solution, the sulphate ions content in calcite decreases significantly while in aragonite it remains constant. A simple model that describes the incorporation of sulphate into the calcite lattice was proposed by Kontrec et al. [102]. In this model, a carbonate group can be replaced by sulphate. Three oxygen atoms of the tetrahedral sulphate structure, which occupies an area of 256.4 $Å^2$, are accommodated in place of the planar carbonate group occupying a smaller surface area of 212.1 $Å^2$. Sulfur and the fourth oxygen of the sulphate group are outside the carbonate plane. As a result, this substitution causes the crystalline lattice of calcite to be distorted. The distortion on the c-axis is more important than on the other two axes (a and b). A comparable model was proposed for the incorporation of sulphate ions into the orthorhombic structure of aragonite [50, 51]. In addition, sulphate ions can be adsorbed on the $CaCO_3$ surfaces [50]. This adsorption was explained by a mechanism involving a change in the surface charge of $CaCO_3$ and a contribution of the Ca^2 ions to the release of carbonate ions from the surface of $CaCO_3$. Indeed, Strand et al. [103] showed that the negatively charged carbonate ions are released from the surface of $CaCO_3$ (desorption) which leads to a positively charged surface. It should be noted that the causes of this desorption have not been mentioned but it was shown that, using a measure of zeta potential, sulphate and calcium ions generate the properties of the $CaCO_3$/water interface. The sulphate ions are adsorbed on this surface and the released carbonate ions react with the Ca^{2+} ions to form $CaCO_3$ on this same surface (**Figure 7**). This adsorption increases with temperature and/or Ca^{2+} ion concentration [103].

The inhibitory effectiveness of iron ions on $CaCO_3$ nucleation, and growth was generally attributed to that (i) iron ions are blocking the growth sites of $CaCO_3$ by incorporating the growing layers, which leads to a strained lattice where calcium ions no longer fit [72] (ii) the formation of colloidal iron oxides, i.e., Fe_2O_3 [104] (iii) iron precipitates mainly in the form of siderite, which acts as heterogeneous sites for $CaCO_3$ precipitation and improves the scale inhibition effectiveness.

The formation of iron oxides and their interactions with $CaCO_3$ remain controversial. For example, experiments were conducted by Herzog et al. [73] on various forms of ferric hydroxide to test if a form is effective for the heterogeneous nucleation of $CaCO_3$. They found that the tested forms of iron hydroxides or

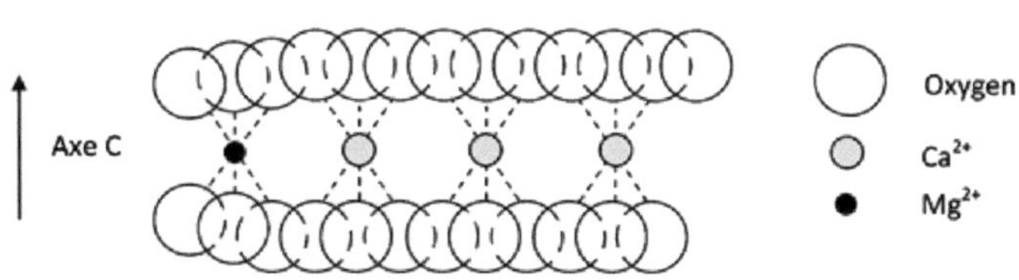

Figure 6.
Calcite lattice distortion due to the incorporation of Mg^{2+}.

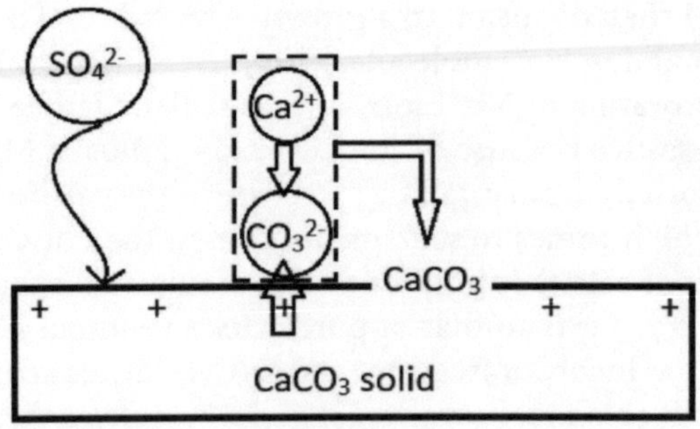

Figure 7.
Schematic illustration of the adsorption of sulphate ions on the surfaces of $CaCO_3$.

hydrated iron oxides did not have any effect on $CaCO_3$ nucleation and growth. This disagrees with the results of Mejri et al. [105]. In fact, they studied the $CaCO_3$ precipitation in 2 mM $Ca(HCO_3)_2$ solution in the presence of iron ions concentrations between 5 and 20 mg L^{-1}. They showed that iron ions addition promotes $CaCO_3$ precipitation. This was attributed to the formation of iron hydroxide before the onset of $CaCO_3$ precipitation which plays a role of seed and initiates the $CaCO_3$ nucleation. Coetzee et al. [104] showed that the addition of iron had a small effect on the crystal morphology but reduced the induction times. The enhanced nucleation rates were explained by the formation of hematite Fe_2O_3 colloids that act as seed crystals and increase the heterogeneous nucleation process. What is more, it was shown that the presence of Fe^{2+} ions in contact with calcite surfaces accelerates calcite dissolution sites during its oxidation and co-precipitation onto calcite surfaces as ferric iron hydroxide, which was rapidly transformed into FeOOH nanoparticles [61]. Thus, the water being treated can be softened by iron precipitation and the final product can be reused as mineral filler powders or as a pigment for industrial applications.

The formation of siderite $FeCO_3$ on the $CaCO_3$ surfaces blocks growth sites and/ or forms a protection layer. This layer retards the $CaCO_3$ precipitation and decreased the corrosion rate of mild steel in simulated saline aquifer environments [106]. The formation of siderite, before the onset $CaCO_3$ precipitation, was experimentally observed by scanning electron microscopy [16]. $FeCO_3$ was formed at relatively low supersaurations and serves as a seed for heteronucleation of vaterite. At high supersaturations, however, $FeCO_3$ either does not form or has stability fields too small to remarkably affect the $CaCO_3$ precipitation. In fact, it was shown that $CaCO_3$ crystallizes in higher amounts than $FeCO_3$ in a solution supersaturated with respect to both $FeCO_3$ and $CaCO_3$ [107], and $FeCO_3$ precipitation is not probable, at high supersaturations, because calcium ions increase the solubility of $FeCO_3$ decreasing, therefore, its precipitation rate [107, 108]. It was shown that, when the carbonate system presents relatively high concentrations, iron solubility is controlled by $FeCO_3$ rather than by $Fe(OH)_2$ in CO_2 saturated solutions [109]. This was confirmed by Korchef's work [16]. Indeed, according to Korchef's mechanism [16], $FeCO_3$ forms first from amorphous iron hydroxide, i.e., consistent with the so-called nonclassical nucleation theory [110] and then serves as a template for heteronucleation of $CaCO_3$. When the concentration of the iron ions increases, the amount of $FeCO_3$ formed increases, which increases the number of heterogeneous sites and enhances the inhibition effectiveness of scale. Then, $FeCO_3$ could be incorporated in a solid solution $Fe_xCa_{1-x}CO_3$ $(0 \leq x \leq 1)$ by a solid–solid transition (**Figure 8**). Indeed, the solids $CaCO_3$ and $FeCO_3$ are isostructural, and their constituent cations (Ca^{2+} and Fe^{2+}) can coexist, as a mixed metal carbonate, in the substitutional solid solution $Fe_xCa_{1-x}CO_3$ [108]. The $FeCO_3$ dissolution was recently reported by Rizzo et al. [111] who studied the effect of $CaCO_3$ precipitation on the corrosion of carbon steel, covered by a protective $FeCO_3$ layer. The authors showed that $CaCO_3$ precipitation lead to an undersaturated solution with respect to $FeCO_3$ followed by dissolution of the protective $FeCO_3$ layer. The partial substitution of Ca^{2+} by Fe^{2+} suppresses the transformation of vaterite into aragonite which results in the decrease of the aragonite amount with increasing iron ions

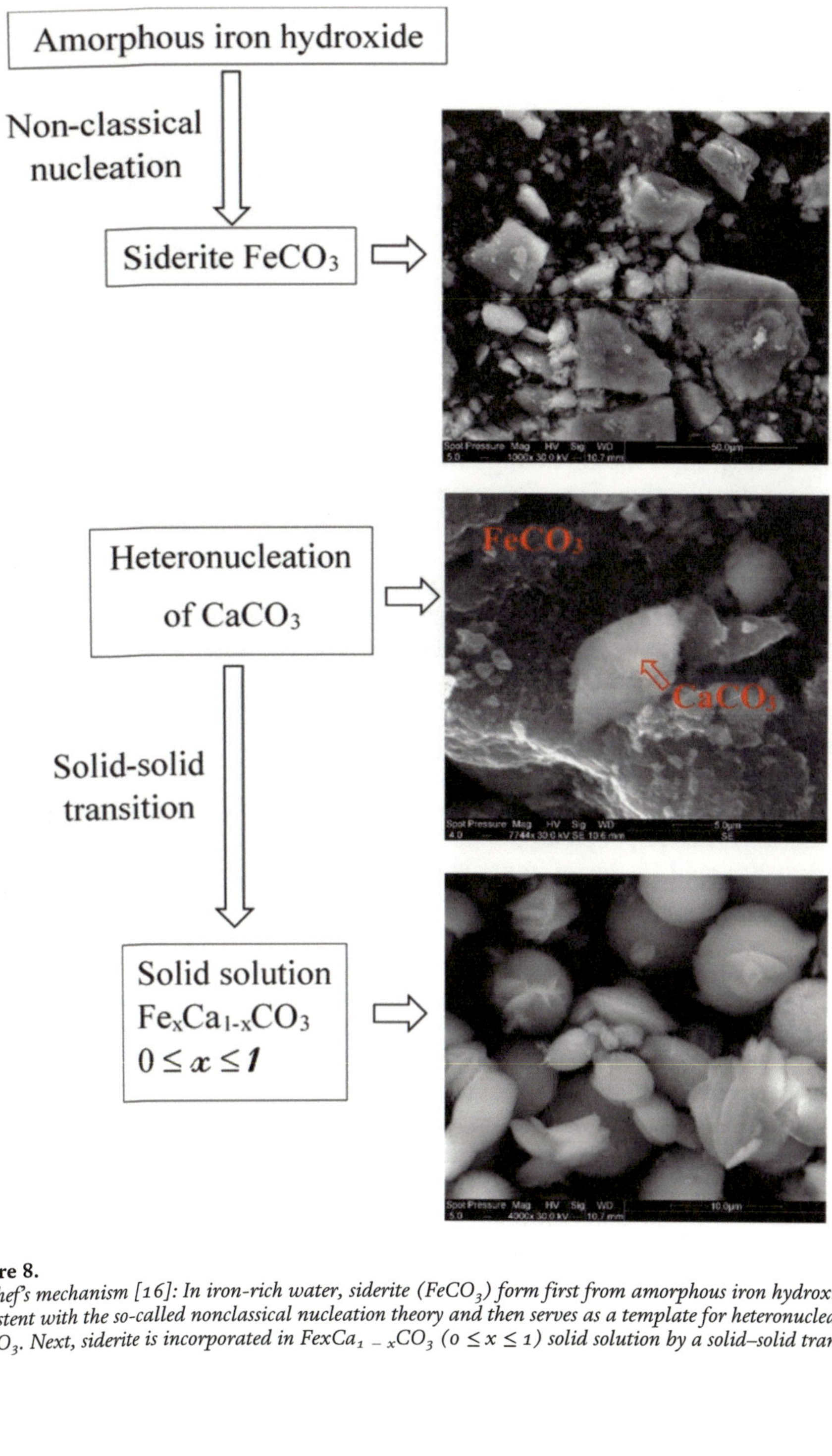

Figure 8.
Korchef's mechanism [16]: In iron-rich water, siderite ($FeCO_3$) form first from amorphous iron hydroxide, consistent with the so-called nonclassical nucleation theory and then serves as a template for heteronucleation of $CaCO_3$. Next, siderite is incorporated in $FexCa_{1-x}CO_3$ ($0 \leq x \leq 1$) solid solution by a solid–solid transition.

concentration. Comparable results were found for Fe(II), [75] Fe(III), [74, 76] and Zn(II) [112]. In fact, Zn(II) precipitated mainly as $ZnCO_3$ which, like siderite, acts as heterogeneous sites in the bulk of solution and decreased the number of crystals on the surface. As a result, scale is reduced [112].

It is worth noticing that, since $FeCO_3$ precipitates at lower pH values than $CaCO_3$ [16], maintaining the solution pH at relatively low values (<7) allowed obtaining only $FeCO_3$. $FeCO_3$ can form protective film on different steel parts of industrial processes which, reduces or completely prevents their corrosion. This reduces remarkably the treatment costs and increases the equipment life.

7. Scale inhibition

Al-Hamza et al. [113] studied the inhibition effect of poly(acrylic acid) (PAA) with different end groups and molar masses on the formation of $CaCO_3$ scale at low and high temperatures. They showed that the inhibition of $CaCO_3$ precipitation was affected by the hydrophobicity of the end groups of PAA. The best inhibition was found for PAA with hydrophobic end groups of moderate size (6–10 carbons). Also, the scale inhibitors stabilize the less thermodynamically stable polymorph (vaterite) to a degree proportional to their ability to inhibit precipitation. At room temperature, the lowest molar mass of PAA with hydrophilic end group showed good efficiency in the inhibition of $CaCO_3$ precipitation making it suitable to be used as a scale inhibitor in reverse osmosis desalination. However, the lowest molar mass PAA with end groups of moderate hydrophobicity are more suitable as scale inhibitors in MSF desalination. The effectiveness of the inhibitors declined with increasing temperature. Comparable results were found by Li et al. [114] who studied the scale inhibition effect of six varieties of commercial scale inhibitors in highly saline conditions at high temperature. Indeed, they showed that the inhibition efficiency was influenced by scale inhibitor dosage, temperature, heating time and pH. The best scale inhibitor (SQ-1211) could effectively retard scaling at high temperature and when the concentration of Ca^{2+} was 1600 mg L^{-1}, the scale inhibition rate reached 90.7% at 80°C and pH 8. The crystal structure of $CaCO_3$ changed from calcite to aragonite. Xu et al. [115] studied the inhibition of $CaCO_3$ scaling on stainless steel surfaces using sodium carboxymethyl cellulose (SCMC). They showed that SCMC exhibited a promising performance of scaling inhibition, and the inhibition efficiency increased with increasing SCMC concentration starting from 50 to 200 mg L^{-1}. For example, the inhibition efficiency reached 93.2% for SCMC concentration of 200 mg L^{-1}. The inhibition effect was explained by the formation of a protective film on the stainless-steel surfaces by adsorption of the constituent of SCMC in the presence of SCMC, preventing the deposition of $CaCO_3$. Zuho et al. [116] investigated the feasibility of electrochemical methods to study the scale inhibition performance of 1-hydroxyethylidene-1,1-diphosphonic acid (HEDP) and 2-phosphonobutane-1,2,4-tricarboxylic acid (PBTCA), polyacrylic acid (PAA) on titanium surfaces. They showed that, for electrochemically deposited $CaCO_3$ scale, PBTCA, HEDP and PAA have great inhibitory effects and the order of the scale inhibition efficiency of the three inhibitors is PBTCA > HEDP > PAA. The deposition of $CaCO_3$ on the surface of TA1 metal mainly results in dendritic aragonite crystals and when scale inhibitors were added, the aragonite crystal polymorph was gradually modified to vaterite. Yu et al. [117] designed and prepared phosphorus-free and biodegradable scale inhibitors. In fact, two series of starch-*graft*-poly(acrylic acid) (St-*g*-PAA) samples with different grafting ratios and grafted-chain distributions, that are, the number and length of grafted PAA chains on the starch backbone were elaborated. They showed that St-*g*-PAA with relatively low grafting ratio (typically ≤97%) inhibited scale formation more effectively than samples with similar grafted-chain distributions. However, under the similar grafting ratios, samples with higher number of branched chains with shorter grafted chains showed better scale efficiency.

More interesting, Korchef and Touabi [17] showed that high amounts of precipitates were obtained with increasing the initial solution pH, and even when formed in the bulk solution, these precipitates can agglomerate and block up pipes pumps and conducts of water in water treatment devices leading to the complete blockage of fluid cooling. For solution pH lower than 7–7.5, no precipitation of $CaCO_3$ was detected [17]. For that reason, the solution pH should be maintained at low values (pH below 7). This can be easily achieved by avoiding CO_2 repelling from the solution. This allows reducing or completely preventing scale in different water treatment processes and favorably affects the water treatment costs and increases the equipment life. In practice, this is achievable by avoiding the contact between water and the atmospheric air. For example, in the desalination units using the solar multiple condensation evaporation cycle (SMCEC) principle, the supersaturation of the feed water increases remarkably with the number of the condensation–evaporation cycles [15] and accordingly, it is expected that iron ions do not inhibit the formation of $CaCO_3$ as scale in the SMCEC desalination units. In that case, it is mandatory to regulate the pH of feed water below 7 to prevent scale formation.

8. Conclusion

Water desalination processes such as reverse osmosis, electrodialysis and distillation are confronted with scale formation which causes enormous energy losses and shortens the equipment life. In most cases, scales are made of $CaCO_3$ which possesses six polymorphs. Environmental conditions play effective roles in the nucleation and growth kinetics of $CaCO_3$ as well as the morphology and polymorphism of the obtained precipitates. In the present chapter, an overview on the effect of operating conditions such as supersaturation, temperature, and pH, on the crystal growth and microstructure of $CaCO_3$ is given. Additionally, I put special emphasis on the effect of foreign ions, especially magnesium, sulphate, chloride, and iron ions. Also, the mechanisms by which these ions affect the crystal growth of $CaCO_3$ were pointed out. The experimental parameters should operate together and control the growth rate and polymorphism of $CaCO_3$. Throughout the present chapter, it was reported that:

- the unstable amorphous $CaCO_3$ is firstly formed. Then, it transformed into crystalline phases which are namely vaterite, aragonite and calcite. The vaterite can be transformed gradually into calcite or aragonite following two steps which are the dissolution of vaterite and the growth of calcite or aragonite. The dissolution of vaterite is affected by the presence of foreign ions, i.e., magnesium, sulphate, and irons ions. The two hydrated crystalline forms of calcium carbonate, calcium carbonate hexahydrate $CaCO_3{\cdot}6H_2O$ and calcium carbonate monohydrate $CaCO_3{\cdot}H_2O$, are more stable than the amorphous $CaCO_3$.

- the crystal growth rate of $CaCO_3$ increases with increasing the solution supersaturation.

- the increase of temperature results to the decrease of the induction time and to the increase of the growth rate of $CaCO_3$ and, also the amount precipitated. At room temperature, calcite is the most predominant phase. At temperatures higher than 50°C, $CaCO_3$ scale precipitates as needles-like aragonite crystals.

- the increase of the initial solution pH accelerates the $CaCO_3$ precipitation. For solution pH lower than 7–7.5, no precipitation of $CaCO_3$ was detected. At pH lower than 8, vaterite was obtained. Pure aragonite needles were formed at pH 10 and a mixture of calcite and aragonite was obtained when increasing the pH from 11 to 12. At high pH values, typically greater than 12, calcite was favored.

- both sulphate and magnesium ions inhibit $CaCO_3$ crystallization, even though some discrepancies on the effect of sulphate ions were revealed.

- iron ions inhibit the $CaCO_3$ growth. The inhibiting effectiveness depends strongly on the solution supersaturation. At relatively low supersaturations, the addition of iron ions retards the $CaCO_3$ precipitation crystals. Past the onset of nucleation, iron ions increased the $CaCO_3$ growth rate and most of $CaCO_3$ amounts precipitated in the bulk solution rather than on the cell walls. This results in the decrease of scaling defined as the precipitation on the cell walls. At high supersaturations, the inhibition effectiveness of iron is small. In that case, the inhibition effectiveness can be enhanced by controlling the operating parameters, i.e., lowering the solution pH, increasing the iron concentration, and/or lowering the solution supersaturation.

- the control of the operating parameters and the water composition allows completely inhibiting or, at least, reducing scale formation in various industrial devices such as desalination units and heat exchangers.

- in practice, the solution pH should be maintained at low values (below 7) by avoiding CO_2 removal from the solution which allows reducing or completely preventing scale. This is realizable by preventing the contact between water and the atmospheric air. The economic impact is of greatest importance since this reduces considerably the treatment costs, enhances the equipment life, and allows increased product water recovery.

Acknowledgements

The author would like to express his gratitude to King Khalid University (KKU), Saudi Arabia, for providing administrative and technical support. The author gratefully acknowledged the researchers, engineers, and technicians of the Laboratory of Natural Water Treatment (LabTEN) of the Centre of Researches and Technologies of Waters (CERTE), Technopark of Borj-Cedria (Tunisia), for many years of fruitful collaboration.

Conflict of interest

The author declares no conflict of interest.

Author details

Atef Korchef[1,2]

1 Joint Programs, College of Science, King Khalid University (KKU), Abha, Saudi Arabia

2 LVMU, Centre National de Recherches en Sciences des Matériaux, Technopole de Borj-Cédria, Tunisia

*Address all correspondence to: akorchef@kku.edu.sa

References

[1] Huang Y, Ji Y, Kang Z, Li F, Ge S, Yang DP, et al. Integrating Eggshell-derived $CaCO_3$/MgO Nanocomposites and Chitosan into a Biomimetic Scaffold for Bone Regeneration. Chem. Eng. J. 2020;**395**:125098. DOI: 10.1016/j.cej.2020.125098

[2] Abbaszadeh M, Nasiri M, Riazi M. Experimental investigation of the impact of rock dissolution on carbonate rock properties in the presence of carbonated water. Environ. Earth Sci. 2016;**75**:791. DOI: 10.1007/s12665-016-5624-3

[3] Kalita JM, Chithambo ML. Thermoluminescence and infrared light stimulated luminescence of limestone ($CaCO_3$) and its dosimetric features. Appl. Radiat. Isot. 2019;**154**: 108888. DOI: 10.1016/j.apradiso.2019.108888

[4] Brečević L, Kralj D. On Calcium Carbonates: from Fundamental Research to Application. Croat. Chem. Acta. 2007;**80**:467-484

[5] Charde SJ, Sonawane SS, Sonawane SH, Navin S. Influence of functionalized calcium carbonate nanofillers on the properties of melt-extruded polycarbonate composites. Chem. Eng. Commun. 2018;**205**: 492-505. DOI: 10.1080/00986445.2017.1404459

[6] Hao BL, Lang Y, Bian DQ, Wang CA. Preparation of near net size porous alumina-calcium aluminate ceramics by gelcasting-pore-forming agent processs. J. Am. Ceram. Soc. 2020;**103**:4602-4610. DOI: 10.1111/jace.17075

[7] Souyris F, Pellequer C, Payrot C, Servera C. Coral, a new biomedical material. Experimental and first clinical investigations on Madreporaria. J. Maxillofac. Surg. 1985;**13**:64-69. DOI: 10.1016/s0301-0503(85)80018-7

[8] Fujioka-Kobayashi M, Tsuru K, Nagai H, Fujisawa K, Kudoh T, Ohe G, et al. Fabrication and evaluation of carbonate apatite-coated calcium carbonate bone substitutes for bone tissue engineering. J. Tissue Eng. Regen. M. 2018;**12**:2077-2087. DOI: 10.1002/term.2742

[9] Woldetsadik AD, Sharma SK, Khapli S, Jagannathan R, Magzoub M. Hierarchically Porous Calcium Carbonate Scaffolds for Bone Tissue Engineering. ACS Biomater. Sci. Eng. 2017;**3**:2457 2469. DOI: 10.1021/acsbiomaterials.7b00301

[10] Elsayed H, Picicco M, Ferroni L, Gardin C, Zavan B, Bernardo E. Novel bioceramics from digital light processing of calcite/acrylate blends and low temperature pyrolysis. Ceram. Int. 2020;**46**:17140-17145. DOI: 10.1016/j.ceramint.2020.03.277

[11] Braye F, Irigaray JL, Jallot E, Oudadesse H, Weber G, Deschamps N, et al. Resorption kinetics of osseous substitute: natural coral and synthetic hydroxyapatite. Biomaterials. 1996;**17**: 1345-1350. DOI: 10.1016/s0142-9612(96)80013-5

[12] Brecevic L. Nielsen AE. Solubility of amorphous calcium carbonate. J. Cryst. Growth. 1989;**98**:504-510. DOI: 101016/0022-0248(89)90168-1

[13] Helalizadeh A, Müller-Steinhagen H, Jamialahmadi M. Mixed salt crystallisation fouling. Chem. Eng. Process. 2000;**39**:29-43. DOI: 101016/S0255-2701(99)00073-2

[14] Mwaba MG, Golriz MR, Gu J. A semi-emperical correlation for crystallisation fouling on heat exchange surfaces. Appl. Therm. Eng. 2006;**26**: 440-450. DOI: 101016/japplthermaleng200505021

[15] Tlili MM, Korchef A, Ben Amor M. Effect of scalant and antiscalant concentrations on fouling in a solar desalination unit. Eng Process. 2007;**46**: 1243-1250. DOI: 101016/jcep200610006

[16] Korchef A. Effect of Iron Ions on the Crystal Growth Kinetics and Microstructure of Calcium Carbonate. Cryst. Growth Des. 2019;**19**:6893-6902. DOI: 101021/acscgd9b00503

[17] Korchef A, Touaibi M. Effect of pH and temperature on calcium carbonate precipitation by CO_2 removal from iron-rich water. Water Environ. J. 2020;**34**: 331-341. DOI: 101111/wej12467

[18] Pytkowicz RM. Rates of inorganic calcium carbonate precipitation. J. Geol. 1965;**73**:196-199. DOI: 10.1086/627056

[19] Cosmo RP, Pereira FAR, Ribeiro DC, Barros WQ, Martins AL. Estimating CO_2 degassing effect on $CaCO_3$ precipitation under oil well conditions. J. Petrol. Sci. Eng. 2019;**181**: 106207. DOI: 10.1016/j.petrol.2019.106207

[20] Hasan BO. Galvanic corrosion of aluminum–steel under two-phase flow dispersion conditions of CO_2 gas in $CaCO_3$ solution. J. Petrol. Sci. Eng. 2015;**133**: 76-84. DOI: 10.1016/j.petrol.2015.05.007

[21] Rizzo R, Gupta S, Rogowska M, Ambat R. Corrosion of carbon steel under CO2 conditions: Effect of CaCO3 precipitation on the stability of the $FeCO_3$ protective layer. Corros. Sci. 2020;**162**:108214. DOI: 10.1016/j.corsci.2019.108214

[22] Gopinath CS, Hegde SG, Ramaswamy AV, Mahapatra S. Photoemission studies of polymorphic $CaCO_3$ materials. Mater. Res. Bull. 2002; **37**:1323-1332. DOI: 10.1016/S0025-5408(02)00763-8

[23] Han YS, Hadiko G, Fuji M, Takahashi M. Factors affecting the phase and morphology of $CaCO_3$ prepared by a bubbling method. J. Eur. Ceramic Soc. 2006;**26**:843-847. DOI: 10.1016/j.jeurceramsoc.2005.07.050

[24] Blue CR, Giuffre A, Mergelsberg S, Han N, DeYoreo JJ, Dove PM. Chemical and physical controls on the transformation of amorphous calcium carbonate into crystalline $CaCO_3$ polymorphs. Geochim. Cosmochim. Acta. 2017;**196**:179-196. DOI: 10.1016/j.gca.2016.09.004

[25] Miyazaki T, Arii T, Shirosaki Y. Control of crystalline phase and morphology of calcium carbonate by electrolysis: Effects of current and temperature. Ceram. Int. 2019;**45**: 14039-14044. DOI: 10.1016/j.ceramint.2019.04.103

[26] Jimoh OA, Ariffin KS, Hussin HB, Temitope AE. Synthesis of precipitated calcium carbonate: A review. Carbonate Evaporite. 2018;**33**:331-346. DOI: 10.1007/s13146-017-0341-x

[27] Plummer LN, Busenberg E. The solubility of calcite, aragonite and vaterite in solutions between 0 and 90°C and an evaluation of the aqueous model for the system $CaCO_3$-CO_2-H_2O. Geochim. Cosmochim. Acta. 1982;**46**: 1011-1040. DOI: 10.1016/0016-7037(82)90056-4

[28] Konopacka-Łyskawa D. Synthesis methods and favorable conditions for spherical vaterite precipitation: A review. Crystals. 2019;**9**:223. DOI: 10.3390/cryst9040223

[29] Akhtar K, Yousafzai S. Morphology control synthesis of nano rods and nano ovals $CaCO_3$ particle systems. J. Disper. Sci. Technol. 2020;**41**:1-15. DOI: 10.1080/01932691.2019.1626248

[30] Rom I, Klas S. Kinetics of $CaCO_3$ precipitation in seeded aeration softening of brackish water desalination concentrate. Chemosphere. 2020;**260**:

127527. DOI: 10.1016/j.chemosphere.2020.127527

[31] Haaring R, Kumar N, Bosma D, Poltorak L, Sudhölter EJR. Electrochemically Assisted deposition of calcite for application in surfactant adsorption studies. Energ. Fuels. 2019; **33**:805-813. DOI: 10.1021/acs.energyfuels.8b03572

[32] Alimi F, Tlili MM, Gabrielli C, Maurin G, Ben Amor M. Effect of a magnetic water treatment on homogeneous and heterogeneous precipitation of calcium carbonate. Water Res. 2006;**40**:1941-1950. DOI: 10.1016/j.watres.2006.03.013

[33] Roques H. Chemical Water Treatment: Principles and Practice. New York: VCH; 1996

[34] Hamer WJ. Theoretical mean activity coefficients of strong electrolytes in aqueous solutions from 0 to 100 °C. US. Natl. Bur. Stand., Ref. Data Ser. 1968;**24**:271

[35] Truesdell AH, Jone BFA. Computer program for calculating chemical equilibria of natural waters. U.S. Geol. Survey J. Res. 1974;**2**:233-–274

[36] Elfil H, Roques H. Prediction of the limit of the metastable zone in the $CaCO_3$-CO_2-H_2O system. AIChE J. 2004;**50**:1908-1916. DOI: 10.1002/aic.10160

[37] Mejri W, Korchef A, Tlili MM, Ben Amor M. Effects of temperature on precipitation kinetics and microstructure of calcium carbonate in the presence of magnesium and sulphate ions. Desalin. Water Treat. 2014;**52**:4863-4870. DOI: 10.1080/19443994.2013.808813

[38] Söhnel O, MullinJ W. Precipitation of calcium carbonate. J. Cryst. Growth. 1982;**60**:239-250. DOI: 10.1016/0022-0248(82)90095-1

[39] Chen J, Xiang L. Controllable synthesis of calcium carbonate polymorphs at different temperatures. Powder Technol. 2009;**189**:64-69. DOI: 10.1016/j.powtec.2008.06.004

[40] Weiss JCA, Cancel KT, Moser RD, Allison PG, Gore ER, Chandler MQ, et al. Influence of temperature on calcium carbonate polymorph formed from ammonium carbonate and calcium acetate. J. Nanotech Smart Mater. 2014; **1**:1-6

[41] Ruiz-Agudo E, Putnis CV, Rodriguez-Navarro C, Putnis A. Effect of pH on calcite growth at constant aCa^{2+}/ aCO_3^{2-} ratio and supersaturation. Geochim. Cosmochim. Acta. 2011;**75**:284-296. DOI: 10.1016/j.gca.2010.09.034

[42] Ramakrishna C, Thenepalli T, Huh JH, Ahn JW. Precipitated calcium carbonate synthesis by simultaneous injection to produce nanowhisker aragonite. J. Korean Ceram. Soc. 2016; **53**:222-226. DOI: 10.4191/kcers.2016.53.2.222

[43] Chang R, Kim S, Lee S, Choi S, Kim M, Park Y. Calcium carbonate precipitation for CO_2 storage and utilization: A review of the carbonate crystallization and polymorphism. Front Energy Res. 2017;**5**:1-–12. DOI: 10.3389/fenrg.2017.00017

[44] Kralj D, Brečević L, Nielsen AE. Vaterite growth and dissolution in aqueous solution. I. Kinetics of crystal growth. J. Cryst. Growth. 1990;**104**:793-800. DOI: 10.1016/0022-0248(90)90104-S

[45] Tai CY, Chen FB. Polymorphism of $CaCO_3$ precipitated in a constant-composition environment. AIChE J. 1998;**44**:1790-1798. DOI: 10.1002/aic.690440810

[46] Chen J, Xiang L. Controllable synthesis of calcium carbonate polymorphs at different temperatures.

Powder Technol. 2009;**189**:64-69. DOI: 10.1016/j.powtec.2008.06.004

[47] Gabrielli C, Maurin G, Poindessous G, Rosset R. Nucleation and growth of calcium carbonate by an electrochemical scaling process. J. Cryst. Growth. 1999;**200**: 236-250. DOI: 10.1016/S0022-0248(98) 01261-5

[48] Devos O, Gabrielli C, Tlili MM, Tribollet B. Nucleation-Growth process of scale electrodeposition: Influence of the supersaturation. Electrochem. Soc. 2003;**150**:494-501

[49] Ben Amor M, Zgolli D, Tlili MM, Manzola SA. Influence of water hardness, substrate nature and temperature on heterogeneous calcium carbonate nucleation. Desalination. 2004;**166**:79-84. DOI: 10.1016/j.desal.2004.06.061

[50] Karoui H, Korchef A, Tlili MM, Mosrati H, Gil O, Mosrati R, et al. Effects of Mg^{2+}, Ca^{2+} and SO_4^{2-} ions on precipitation kinetics and microstructure of aragonite. Ann. Chim. Sci. Mater. 2008;**33**:123-134

[51] Fernández-Díaz L, Fernández-González Á, Prieto M. The role of sulfate groups in controlling $CaCO_3$ polymorphism. Geochim. Cosmochim. Acta. 2010;**74**:6064-6076. DOI: 10.1016/j.gca.2010.08.010

[52] Liu X, Li K, Wu C, Li Z, Wu B, Duan X, et al. Influence of copper (II) on biomineralization of CaCO3 and preparation of micron pearl-like biomimetic CaCO3. Ceram. Int. 2019; **45**:14354-14359. DOI: 10.1016/j.ceramint.2019.04.150

[53] Mucci A, Morse JW. The incorporation of Mg^{2+} and Sr^{2+} into calcite overgrowths. Influence of growth rate and solution composition. Geochim. Cosmochim Acta. 1983;**4**: 7217-7233. DOI: 0016-7037/83/020217-17503.00/0

[54] Busenberg E, Plummer LN. Kinetic and thermodynamic factors controlling the distribution of SO_4^{2-} and Na^+ in calcites and selected aragonites. Geochim. Cosmochim. Acta. 1985;**49**: 713-725. DOI: 10.1016/0016-7037(85) 90166-8

[55] Hu Z, Deng Y. Supersaturation control in aragonite synthesis using sparingly soluble calcium sulfate as reactants. J. Colloid Interf. Sci. 2003; **266**:359-365. DOI: 10.1016/S0021-9797 (03)00699-4

[56] Chen T, Neville A, Yuan M. Influence of Mg2+ on CaCO3 formation-bulk precipitation and surface deposition. Chem. Eng. Sci. 2006;**61**: 5318-—5327. DOI: 10.1016/j.ces.2006.04.007

[57] Vavouraki AI, Putnis CV, Putnis A, Koutsoukos PG. An atomic force microscopy study of the growth of calcite in the presence of sodium sulfate. Chem. Geol. 2008;**253**:243-251. DOI: 10.1016/j.chemgeo.2008.05.013

[58] Konrad F, Purgstaller B, Gallien F, Mavromatis V, Gane P, Dietzel M. Influence of aqueous Mg concentration on the transformation of amorphous calcium carbonate. J. Cryst. Growth. 2018;**498**:381-390. DOI: 10.1016/j.jcrysgro.2018.07.018

[59] Cuesta Mayorga I, Astilleros JM, Fernandez-Diaz L. Precipitation of $CaCO_3$ polymorphs from aqueous solutions: The role of pH and sulfate groups. Minerals. 2019;**9**:178. DOI: 10.3390/min9030178

[60] Ko M, Laycock NJ, Ingham B, Williams DE. In situ synchrotron X-ray diffraction studies of CO2 corrosion of carbon steel with scale inhibitors ATMPA and PEI at 80°C. Corros. 2012; **68**:1085-1093. DOI: 10.5006/0657

[61] Hamdouni A, Montes-Hernandez G, Tlili MM, Findling N, Renard F, Putnis CV. Removal of Fe(II) from groundwater via aqueous portlandite carbonation and calcite-solution interactions. Chem. Eng. J. 2016;**283**: 404-411. DOI: 10.5006/0657

[62] Korchef A, Kerkeni I, Ben Amor M, Galland S, Persin F. Iron removal from aqueous solution by oxidation, precipitation and ultrafi ltration. Desalin. Water Treat. 2009;**9**:1-8. DOI: 10.5004/dwt.2009.745

[63] EU. Drinking Water Directive; Council Directive 98/83/EC of 3 November 1998. Official Journal of the European Communities L330/32 from 5.12.1998. Available at: http://eurlex.europa.eu/LexUriServ/site/en/oj/1998/l_330/l_33019981205en00320054.pdf

[64] Tlili MM, Ben Amor M, Gabrielli C, Joiret S, Maurin G, Rousseau P. Study of electrochemical deposition of $CaCO_3$ by in situ raman spectroscopy: II. Influence of the solution composition. J. Electrochem. Soc. 2006;**150**:485-493

[65] Tlili MM, Ben Amor M, Gabrielli C, Joiret S, Maurin G. On the initial stages of calcium carbonate precipitation. Eur. J. Water Qual. 2006;**37**:89-108

[66] Paquette J, Vali H, Mucci A. TEM study of Pt-C replicas of calcite overgrowths precipitated from electrolyte solutions. Geochim. Cosmochim. Acta. 1996;**60**:4689-4701. DOI: 10.1016/S0016-7037(96)00270-0

[67] Morse WJ, Arvidson RS, Luttge A. Calcium carbonate formation and dissolution. Chem. Rev. 2007;**107**: 342-381. DOI: 10.1021/cr050358j

[68] Rushdi AI, Pytkowicz RM, Suess E, Chen CT. The effect of magnesium-to-calcium ratios in artificial seawater at different ionic products, upon the induction time and mineralogy of calcium carbonate: A laboratory study. Geol. Rdsch. 1992;**81**:751-758

[69] Reddy MM. Effect of magnesium ion on calcium carbonate nucleation and crystal growth in dilute aqueous solutions at 25° celsius, studies in diageneses, U.S. Geological survey, Denver, Bulletin. 1986; 1578:169–182.

[70] Chen T, Neville A, Yuan M. Assessing the effect of Mg^{2+} on $CaCO_3$ scale formation–bulk precipitation and surface deposition. J. Cryst. Growth. 2005;**275**:1341-1347. DOI: 10.1016/j.jcrysgro.2004.11.169

[71] Kelland MA. Effect of various cations on the formation of calcium carbonate and barium sulphate scale with and without scale inhibitors. Ind. Eng. Chem. Res. 2011;**50**:5852-5861. DOI: 10.1021/ie2003494

[72] Di Lorenzo F, Burgos-Cara A, Ruiz-Agudo E, Putnis CV, Prieto M. Effect of ferrous iron on the nucleation and growth of $CaCO_3$ in slightly basic aqueous solutions. CrystEngComm. 2017;**19**: 447-460. DOI: 10.1039/c6ce02290a

[73] Herzog RE, Shi Q, Patil JN, Katz JL. Magnetic water treatment: The effect of iron on calcium carbonate nucleation and growth. Langmuir. 1989;**5**:861-867. DOI: 10.1021/la00087a048

[74] Katz JL, Reik MR, Herzog RE, Parsiegla KI. Calcite growth inhibition by iron. Langmuir. 1993;**9**:1423-1430. DOI: 10.1021/la00029a043

[75] Pernot B, Euvrard M, Simon P. Effect of iron and manganese on the scaling potentiality of *water*. J. Water SRT – Aqua. 1998;**47**:21-29. DOI: 10.2166/aqua.1998.0004

[76] Takasaki S, Parsiegla KI, Katz JL. Calcite growth and the inhibitory effect of iron(III). J. Cryst. Growth. 1994;**143**: 261-268. DOI: 10.1016/0022-0248(94)90066-3

[77] Macadam J, Parsons SA. Calcium carbonate scale and control, effect of material and inhibitors. Water Sci. Technol. 2004;**49**:153-159. DOI: 10.2166/wst.2004.0112

[78] Ruiz-Agudo E, Putnis CV, Rodrigues-Navarro C, Putnis A. Effect of pH on calcite growth at constant aCa^{2+}/aCO_3^{2-} ratio and supersaturation. Geochim. Cosmochim. Acta. 2011;**75**: 284-296. DOI: 10.1016/j.gca.2010.09.034

[79] Elgqulst B, Wedborg M. Stability of ion pairs from gypsum solubility, degree of ion pair formation between the major constituents. Mar. Chem. 1975;**3**:215-225. DOI: 10.1016/0304-4203(75)90003-1

[80] Sucha L, Cadek J, Hrabek K, Vesely J. The stability of the chloro complexes of magnesium and of the alkaline earth metals at elevated temperature. Collect. Czech. Chem. Commun. 1975;**40**:2020-2024. DOI: 10.1135/cccc19752020

[81] Johnson KS, Pytkowicz RM. Ion association with H+, Na+, K+, Ca2+, and Mg2+ in aqueous solutions at 25 degrees C. Am. J. Sci. 1978;**278**:1428-1447. DOI: 10.2475/ajs.278.10.1428

[82] Majer V, Stulik K. A study of the stability of alkaline earth metal complexes with fluoride and chloride ions at various temperatures by potentiometry with ion selective electrodes. Talanta. 1982;**29**:145-148. DOI: 10.1016/0039-9140(82)80039-8

[83] Allakhverdov GR. Calculation of the formation constants of singly charged complex ions of bivalent metals in solution. Russ. J. Phys. Chem. 1985;**59**: 39-41

[84] Williams-Jones AE, Seward TM. The stability of calcium chloride ion pairs in aqueous solutions at temperatures between 100 and 360°C. Geochim. Cosmochim. Acta. 1989;**53**: 313-318. DOI: 10.1016/0016-7037(89)90383-9

[85] Takita Y, Eto M, Sugihara H, Nagaoka K. Promotion mechanism of co-existing NaCl in the synthesis of $CaCO_3$. Mater. Lett. 2007;**61**:3083-3085. DOI: 10.1016/j.matlet.2006.11.005

[86] Ukrainczyk M, Gredicak M, Jeric I, Kralj D. Interactions of salicylic acid derivatives with calcite crystals. J. Colloid Interface Sci. 2012;**365**:296-307. DOI: 10.1016/j.jcis.2011.09.009

[87] Natsi PD, Rokidi SG, Koutsoukos PG. Precipitation of calcium carbonate ($CaCO_3$) in water-monoethylene glycol solutions. Ind. Eng. Chem. Res. 2019;**58**:4732-4743. DOI: 10.1021/acs.iecr.8b04180

[88] Oral CM, Kapusuz D, Ercan B. Enhanced vaterite and aragonite crystallization at controlled ethylene glycol concentrations. Sakarya Univ. J. Sci. 2019;23:129–138. DOI: 10.16984/saufenbilder.433985

[89] Zhang D, Lin Q, Xue N, Zhu P, Wang Z, Wang W, et al. The kinetics, thermodynamics and mineral crystallography of $CaCO_3$ precipitation by dissolved organic matter and salinity. Sci. Total Environ. 2019;**673**:546-552. DOI: 10.1016/j.scitotenv.2019.04.138

[90] Džakula BN, Fermani S, Dubinsky Z, Goffredo S, Falini G, Kralj D. In vitro coral biomineralization under relevant aragonite supersaturation conditions. Chem. Eur. J. 2019;**25**:10616-10624. DOI: 10.1002/chem.201900691

[91] Ostwald W. Lehrbuch für allgemeine Chemie, Vol II, Engelman, Leipzig, 1902.

[92] Bamford CH, Tipper CFH. Comprehensive Chemical Kinetics, Vol. 22, Reactions in the Solid State, Elsevier, Amsterdam, 1980.

[93] Cardew PT, Davey RJ, Ruddick AJ. Kinetics of polymorphic solid-state transformations. J. Chem. Soc., Faraday Trans. 2. 1984;**80**:659-668. DOI: 10.1039/F29848000659

[94] Davey RJ, Ruddick AJ, Guy PD, Mitchell B, Maginn SJ, Polywka LA. The IV-III polymorphic phase transition in ammonium nitrate: A unique example of solvent mediation. J. Phys., D: Appl. Phys. 1991;**24**:176-185

[95] Kralj D, Brečević L, Kontrec J. Vaterite growth and dissolution in aqueous solution III. Kinetics of transformation. J. Cryst. Growt. 1997; **177**:248-257. DOI: 10.1016/S0022-0248 (96)01128-1

[96] Han YS, Hadiko G, Fuji M, Takahashi M. Influence of initial CaCl2 concentration on the phase and morphology of $CaCO_3$ prepared by carbonation. J. Mater. Sci. 2006;(14): 4663-4667. DOI: 10.1007/s10853-006-0037-4

[97] Kitano Y, Okumura M, Idogaki M. Incorporation of sodium, chloride and sulfate with calcium carbonate. Geochem Journal. 1975;**9**:75-84. DOI: 10.2343/geochemj.9.75

[98] Menadakis M, Maroulis G, Koutsoukos PG. Incorporation of Mg^{2+}, Sr^{2+}, Ba^{2+} and Zn^{2+} into aragonite and comparison with calcite. J. Math. Chem. 2009;**46**:484-491. DOI: 10.1007/s10910-008-9490-4

[99] Lahann RW. A chemical model for calcite crystal growth and morphology control. J. Sediment. Petrol. 1978;**48**: 337-344. DOI: 10.1306/212F746E-2B24-11D7-8648000102C1865D

[100] Folk RL. The natural history of crystalline calcium carbonate: effect of magnesium content and salinity. J. Sediment. Petrol. 1974;**44**:40-35. DOI: 10.1306/74D72973-2B21-11D7-8648000102C1865D

[101] Chave KE, Deffeyes KS, Weyl PK, Garrels RM, Thompson ME. Observations on the solubility of skeletal carbonates in aqueous solutions. Science. 1962;**137**:33-34. DOI: 10.1126/science.137.3523.33

[102] Kontrec J, Kralj D, Brečević L, Falini G, Fermani S, Noethig-Laslo V, et al. Incorporation of inorganic anions in calcite. Eur. J. Inorg. Chem. 2004; (23):4579-4585. DOI: 10.1002/ejic.200400268

[103] Strand S, Hognesen EJ, Ausad T. Wettability alteration of carbonates—Effects of potential determining ions (Ca^{2+} and $SO_4{}^{2-}$) and temperature. Colloids Surf. A: Physicochem. Eng. Aspects. 2006;**275**:1-10. DOI: 10.1016/j.colsurfa.2005.10.061

[104] Koetzee PP, Yacoby M, Howall S, Mubenga S. Scale reduction and scale modification effect induced by Zn and other metal species in physical water treatment. Water SA. 1998;**24**:77-84

[105] Mejri W, Ben Salah I, Tlili MM. Speciation of Fe(II) and Fe(III) effect on CaCO3 crystallization. Cryst. Res. Technol. 2015;**50**:236-243. DOI: 10.1002/crat.201400444

[106] Esmaeely SN, Choi YS, Young D, Nesic S. Effect of Calcium on the formation and protectiveness of iron carbonate layer in CO_2 corrosion. Corrosion. 2013;**69**:912-920. DOI: 10.5006/0942

[107] Alsaiari HA, Kan A, Tomson M. Effect of calcium and iron (II) ions on the precipitation of calcium carbonate and ferrous carbonate. Soc. Petrol. Eng. J. 2010;**15**:294-300. DOI: 10.2118/121553-PA

[108] Mansoori H, Young D, Brown B, Nesic S, Singer M. Effect of $CaCO_3$-saturated solution on CO_2 corrosion of mild steel explored in a system with controlled water chemistry and well-

defined mass transfer conditions. Corros. Sci. 2019;**158**:108078. DOI: 10.1016/j.corsci.2019.07.004

[109] Morgan B, Lahav O. The effect of pH on the kinetics of spontaneous Fe(II) oxidation by O_2 in aqueous solution-basic principles and a simple heuristic description. Chemosphere. 2007;**68**: 2080-2084. DOI: 10.1016/j.chemosphere.2007.02.015

[110] Wang T, Colfen H, Antonietti M. Nonclassical crystallization: Mesocrystals and morphology change of $CaCO_3$ crystals in the presence of a polyelectrolyte additive. J. Am. Chem. Soc. 2005;**127**:3246-3247. DOI: 10.1021/ja045331g

[111] Rizzo R, Gupta S, Rogowska M, Ambat R. Corrosion of carbon steel under CO2 conditions: Effect of $CaCO_3$ on the stability of the $FeCO_3$ protective layer. Corros. Sci. 2019;**162**:108214. DOI: 10.1016/j.corsci.2019.108214

[112] Pernot B, Euvrard M, Remy F, Simon P. Influence of Zn(II) on the crystallisation of calcium carbonate application to scaling mechanisms. J. Water Serv. Res. Technol. Aqua. 1999; **48**:16-23

[113] Al-Hamzah AA, East CP, Doherty WOS, Fellows CM. Inhibition of homogenous formation of calcium carbonate by poly (acrylic acid). The effect of molar mass and end-group functionality. Desalination. 2014;**338**: 93-105. DOI: 10.1016/j.desal.2014.01.020

[114] Li X, Gao B, Yue Q, Ma D, Rong H, Zhao P, et al. Effect of six kinds of scale inhibitors on calcium carbonate precipitation in high salinity wastewater at high temperatures. J. Environ. Sci. 2015;**29**:124-130. DOI: 10.1016/j.jes.2014.09.027

[115] Xu Z, Zhao Y, Wang J, Chang H. Inhibition of calcium carbonate fouling on heat transfer surface using sodium carboxymethyl cellulose. Appl. Therm. Eng. 2019;**148**:1074-1080. DOI: 10.1016/j.applthermaleng.2018.11.088

[116] Zuo Z, Yang W, Zhang K, Chen Y, Li M, Zuo Y, et al. Effect of scale inhibitors on the structure and morphology of $CaCO_3$ crystal electrochemically deposited on TA1 alloy. J. Colloid Interf. Sci. 2020;**562**: 558-566. DOI: 10.1016/j.jcis.2019.11.078

[117] Yu W, Wang Y, Li A, Yang H. Evaluation of the structural morphology of starch- graft -poly(acrylic acid) on its scale-inhibition efficiency. Water Res. 2018;**141**:86-95. DOI: 10.1016/j.watres.2018.04.021

Chapter 3

Cadmium Contents in Biodegradable Films Made from Cassava

Maite Rada-Mendoza, José Luis Arciniegas Herrera and Patricia Vélez Varela

Abstract

A Review of sample preparation and techniques used to determine Cadmium content in flexible films and biodegradable thermoformed products based on cassava. All determinations have been made using atomic absorption spectroscopy. The presence of this element in these matrices can be harmful if the maximum tolerated quantities are not complied by. The presence of Cadmium may be due, among other aspects, to the fact that it is present in the raw material, in industrial discharges or because it is found naturally. Its determination is an important parameter that needs to be considered as a good alternative for packaging.

Keywords: cadmium, thermoformed, flexible films, atomic absorption

1. Background and objectives

1.1 Introduction

The accelerated pace of industrialization, combined with rapid population growth, intensive agricultural techniques, and inappropriate waste management in developing countries have increased the levels of micropollutants such as heavy metals, considered harmful or toxic to living beings [1, 2]. These heavy metals can enter the human body —causing serious damage— via food, water, air, soil, skin absorption, polluting emissions, anthropogenic sources (treated sewage discharges, mining operations), contact with industrial and agricultural products such as pesticide formulations, urban traffic, contamination from chemical fertilizers, and irrigation with poor quality water [3–8].

The migration of heavy metals induced by substances in contact with food is given by the negative interaction between packaging and food. Another risk factor is constituted by the fact that heavy metals are nonbiodegradable and cannot be metabolized and, thus, persist and accumulate in the environment and in organisms over long periods of time [1–3, 9–11].

1.2 Biodegradable thermoformed and flexible films

Thermoformed and flexible films are very popular in the food packaging market given their low cost, large surface area by volume, and outstanding performance across a wide temperature range [12]. Thermoforming is a generic term that encompasses several processing techniques, by which plastic articles can be obtained from flat sheets of different polymers. Thermoformed products are classified into two major categories: permanent or industrial products (shelves for medical or electronic equipment, decorative panels for cars, planes, motorcycles, bathtubs and bathroom fixtures, helmets and seats for boats, and skylights) and disposable products (packaging for medicines, bubble wrap, cups for hot and cold beverages, baking trays, food containers, and clear packaging that is shaped like the product) [13, 14].

The films are defined as thin flexible sheets of synthetic or natural origin, that reach a thickness of 0.01 inches or less. The polymeric matrix that forms the film can be defined as the random arrangement of the chains that make up the structure. Flexible films used in the food industry are currently divided into two groups: synthetic or non-biodegradable materials and biodegradable materials. The manufacture of these products is important, as they break down quickly and easily, without producing residues that cause unfavorable impacts on environmental ecosystems [14].

Biodegradable thermoformed and flexible films can be made of cassava flour (given its high starch content, making it suitable for use in the manufacture of various products in the food industry, as well as for the production of biopolymers) and cassava starch (as it can be converted into a thermoplastic material by interrupting the molecular interactions of the double helix chain, formed by hydrogen bridges between the hydroxyl groups, in the presence of a plasticizer aided by suitable temperature and shear stress [15, 16], and by adding matter such as fique fiber, gelatin, poly(butylene adipate-coterephthalate), polylactic acid, glycerol, plasticizer, green composites, cellulose, chitosan, clay, pullulan, natural extracts, poly(vinyl alcohol), and kaolin [7, 8, 17–35].

Industries are currently striving to improve their products by using natural and renewable sources to store, package, and wrap food products [7, 8, 29, 32, 36]. These include thermoformed and flexible films (biopolymers) obtained from agricultural sources (cassava), and constitute a new and environmentally friendly industrial alternative (composting) given their rapid and easy biodegrading processes. Their main purpose is the replacement of regularly used petroleum-derived plastic polymers and their associated waste [8, 18, 20, 22, 23, 26, 31, 37, 38]. A common production method for thermoforming is the compression molding technique (**Figure 1**), in which the material is placed into an open mold to which pressure and heat are applied.

Figure 1.
Compression molding machine used for thermoforming.

Single screw extrusion —a continuous process that forms the plastic material— is used to produce biodegradable flexible films, whereby turning the screw and applying heat pushes the material along while melting it [13, 21, 27, 32], turning it into a viscous material. Where the nozzle is positioned at the end of the cylinder, there is a hole shaped according to the use required (for films, the extrusion nozzle is round). From the extrusion hopper comes the plasticized mass, which is stretched, smoothed, and rolled by the rollers (**Figure 2**) [39].

The casting technique, using native and ozonated cassava starch, glycerol as the plasticizer, and water as the solvent is used for the elaboration of biodegradable films from cassava starch modified by ozone at different levels [40].

1.3 A case study of heavy metal pollution: cadmium

Biodegradable polymers may lead to heavy metal contamination during their manufacture if the raw material used to process them is contaminated (cassava flour and starch, fique fiber, polylactic acid, glycerin, etc.). They may also be subject to cross-contamination during production, for example, via petroleum- and nonpetroleum-associated activities, from containers used for cooking and storage, during drying, and from contaminated utensils or water [4, 41, 42].

Thus, tracking the manufacturing process is essential in determining the presence of toxic heavy metals in the biopolymers, in turn, to safeguard public health by limiting exposure [1, 9].

Among these metals, cadmium (Cd) is considered toxic to plants and animals; it is widespread in the atmosphere, soils, and water, and is a serious health hazard, affecting the gastrointestinal, cardiovascular, musculoskeletal, nervous, renal, and reproductive systems. Long-term exposure may cause mitochondrial damage and possible death. Cd bioaccumulates in individual organisms [4, 5, 43–46] and is likely to continue to do so due to the future use of biodegradable polymers in the packing and packaging of dry foods and other degradable products, as it enters the food chain.

The US Department of Health and Human Services has insisted that an excessive Cd accumulation in humans may be the cause of cancer as Cd and Cd salts are considered a "possible human carcinogen" [4, 44]. Certain plants can also accumulate Cd in their tissues, and levels are even more significant in plants grown in peri-urban areas with soil contaminated by irrigation using wastewater and sewage sludge [10, 47–50].

There are currently no studies reporting Cd found in thermoformed and biodegradable flexible films; however, the literature does report the presence of Cd in cassava tubers' cortex [42], cassava food crops [51], as well as Cd contaminated rice which causes the Itai-itai disease [52] by exposure to Cd-contaminated water

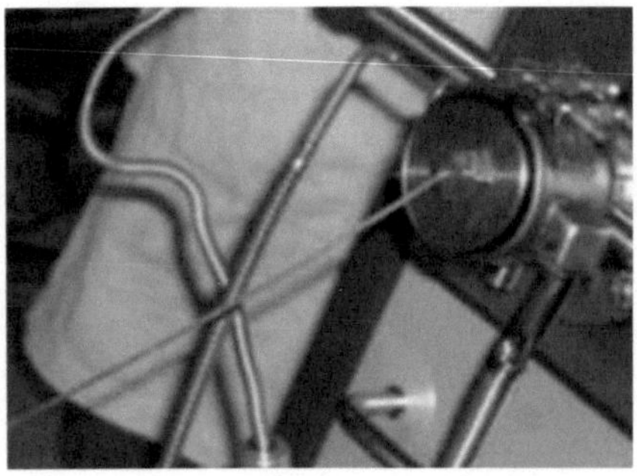

Figure 2.
Film extrusion machine.

used for irrigation or from farms affected by petroleum- and nonpetroleum-related activities. The safety of the materials that come into contact with food is assessed by the amount of substances that migrate into it from the biopolymer and whether or not these meet the conditions established in the legislation on foods.

The legislation limits the migration of toxic pollutants from reaching unacceptable levels and helps to maintain the integrity of foodstuffs, thereby preventing health hazards, contamination, and alternations in food composition and sensory properties [9]. The current Colombian regulation NTC 4096 [53], allows a maximum level of Cd of 1 mg/kg for plasticizers while the Agency for Toxic Substances and Disease Registry [44] estimated that average Cd intake in Americans is 30 mg/day and that only one-tenth of this amount is absorbed into the tissues. Given that thermoformed and flexible film can be used as food containers, their Cd levels must be strictly monitored.

2. Sample preparation and analytical techniques

2.1 Sample pre-treatment

The samples must be treated prior to analysis. The particle size of thermoformed samples are reduced via mechanical methods, starting with manual cracking; followed by maceration (**Figure 3**) to a particle size of less than 1,135 mm; and finally, the removal of water content by placing the particles in an oven at 70 °C for 3 hours [54].

For flexible film samples, all that is required is manual cutting as shown in **Figure 4**, and drying at 50 °C for 45 minutes [54–58].

The samples' moisture content can be determined in triplicate by calculating the weights obtained before and after drying (see Eq. (1)).

$$Moisture\,(\%) = \frac{Moist\,sample\,weight - Dry\,sample\,weight}{Moist\,sample\,weight} * 100 \tag{1}$$

Following drying in a furnace, the moisture values for thermoformed samples were between 3.71% and 5.80% (RSD lower than 3.4%). Biodegradable flexible films revealed higher values of moisture (following 45 mins of drying in a furnace) at 7.81–10.35%, with relative standard deviation (RSD) lower than 1.96% [54].

2.2 Optimization of acid digestion using the reflux system

When an analyte cannot be determined, it must be transformed to a state in which an appropriate identification and quantification technique can be applied.

Figure 3.
Sample: Thermoforming and maceration.

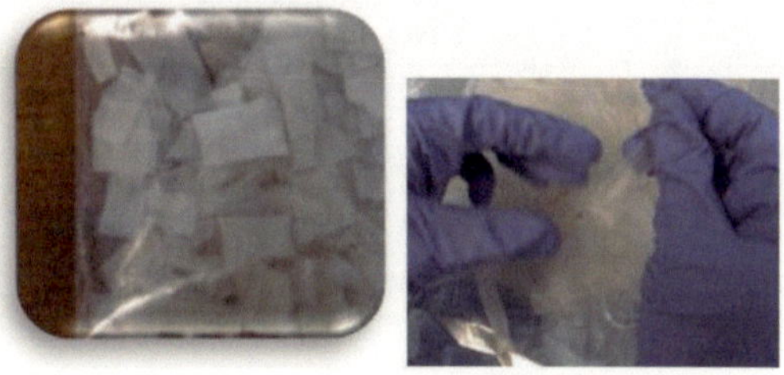

Figure 4.
Sample: Flexible films and cutting.

These transformations are usually dissolutions or digestions, which involve the sample passing from a solid state to a liquid one, using the correct solvent (acids or bases of different strength, oxidizing agents, or enzymes, etc.); in this energetic process, heating and agitation increase the speed of mass transfer, unifying the decomposition and elimination of organic matter, and leaving the trace components of interest (metallic ions) in solution [59, 60].

In this procedure, the organic matter is destroyed by a wet process and the sample is digested (with an acid/oxidant mixture), combining sulfuric, nitric, perchloric acid or hydrogen peroxide, in a system that can be open or refluxed [61];

This technique is generally preferred to dry oxidation (where the sample is heated to 450–500 °C), as the presence of large volumes of acids produces less loss of trace elements through evaporation. However, there is also the danger of elements being lost by evaporation (antimony, arsenic, boron, chromium, tin, germanium, mercury, osmium, rhenium and selenium), although these losses can be prevented by adjusting the conditions (mounting of reflux, temperature control and time) [59, 60].

A wide range of acid digestion procedures are reported in the literature, in which various mixtures of inorganic acids and, in some cases, hydrogen peroxide have been used (HNO_3, HNO_3-H_2SO_4, HNO_3-HF, HNO_3-H_2O_2, HNO_3-HF–H_2O_2, HNO_3-$HClO_4$, y HNO_3-$HClO_4$–H_2O_2) [11, 62–64].

Of the mineral acids, nitric acid offers the best digestion result for all types of samples; however, there is no consensus on the addition of other substances, such as perchloric acid or hydrogen peroxide, to accelerate the process and reduce the volume of nitric acid used [60].

This study employed acid digestion in its sample preparation [51, 54]. Optimizing acid digestion using the reflux system involved the determination of cadmium in a thermoformed sample (HMC-1) and flexible film (SM 707–17 hydrolyzed), taking four absorbance readings and considering the following variables: sample weight, temperature, time, and acid ratio. Each test should be performed in triplicate.

2.2.1 Acid ratio optimization

The optimal acid ratio for the digestion of thermoformed samples is determined experimentally, by varying the amount of mineral acids: HNO_3 at 65% (20 to 5 mL) and $HClO_4$ at 48% (5 to 20 mL), as shown in **Table 1**. Each test should be performed in triplicate.

For flexible films, the optimization was only possible with an acid ratio of 3:1 as this was the best response to the digestion procedure for thermoforming.

Digestion was performed using the sample dissolved in 20 mL of a mixture containing HNO_3 (65%, Merck): perchloric acid (48%, Merck) at a 3:1 ratio [54].

HNO_3:$HClO_4$ Ratio	HNO_3 (mL)	$HClO_4$ (mL)
1:0	20.0	—
3:1	15.0	5.0
1:1	10.0	10.0
1:3	5.0	15.0
0:1	—	20.0

Table 1.
Acid ratio (HNO3 and HClO4) for acid digestion using the reflux system.

2.2.2 Sample weight optimization

This parameter is optimized by varying the sample quantity, which for our study, has been: 0.5000 (± 0.0001) g, 1.0000 (± 0.0001) g and 2.0000 (± 0.0001) g, for thermoforming, and 0.5000 (± 0.0001) g and 1.0000 (± 0.0001) for flexible films, on a dry base.

Digestion was performed using a 1.0 g of the sample dissolved in acid [54].

2.2.3 Temperature and digestion time optimization

The optimal digestion temperature is determined experimentally at three heating temperatures (35, 50 and 70 °C) for both thermoformed samples and flexible films.

For digestion time, 60, 120 and 180 minutes were tested for the thermoformed samples, for the acid ratios shown in **Table 1**, and for flexible films 15, 30, 45, 60 and 120 minutes, for the acid ratio 3:1. After cooling to room temperature, all digested solutions were filtered using a filter crucible, and then stored in polyethylene containers at 4 °C for further analysis by atomic absorption spectrometry.

Thermoformed samples were dried at 70 °C in a furnace for 3 h. Biodegradable flexible films were dried at 50 °C in a furnace for 45 min [54].

Numerous techniques have been used to determine the concentration of heavy metals in different samples, such as X-ray fluorescence spectrometry (XRFS), atomic absorption spectrometry (AAS) and inductively coupled plasma mass spectrometry (ICP-MS). XRFS makes it possible to analyze solid materials without sample pretreatment; however, this advantage is limited by the need to use appropriate certified reference materials for calibration, making it a very expensive technique. In ICP-MS and EAA, dissolved liquid samples are usually required, so the samples have to be previously digested. This procedure can be tedious, time consuming and results in systematic errors due to incomplete extraction or solubility of the analyte [65].

When compared to the flame atomization method, the graphite furnace atomic absorption spectrometry is the most suitable technique to determine elements such as Cd, in trace level concentrations in polymer samples. The latter has many advantages such as its high sensitivity, its limits of detection in the order of micrograms per liter μg/L to ng/L, its tolerance to complex matrices, the fact that it minimizes analyte loss, that it uses few sample volumes, and that it reduces analyst contamination risk [65–69]. With this technique, samples are often introduced in solution form, however, the EAA-HG technique with solid samples has been reported as a simple and fast method for the determination of lead and cadmium in polymer samples [65].

2.3 Statistical treatment

The following statistical tests have to be applied in order to analyze the degree of agreement between the individual data obtained for the standardization of the analytical method:

The Shapiro–Wilk and Levene tests are used to evaluate the normality and homogeneity of the data obtained for repeatability and intermediate precision, with a confidence interval greater than 95%. It is also necessary to determine the standard deviation and the coefficient of variation to determine whether the method is precise.

For linearity, the Pearson correlation coefficient and the coefficient of determination (R2) must be determined, a one-way ANOVA applied, and the t-student test performed to evaluate whether the slope differs significantly from zero.

For samples of thermoformed, flexible films, flour and starch samples, one-way ANOVA needs to be applied in order to determine whether there are any significant differences between them.

For statistical analysis, programs such as: Microsoft Office Excel 2007 and SPSS Statistical Software version 11.5 are used.

3. Fine-tuning of the analytical technique

The analytical technique has to be fine-tuned prior to analysis:

3.1 Determine the stability of the cadmium hollow cathode lamp

To determine the stability of the lamp, the absorbance value of at least three element standards of different concentrations 0.2, 0.6 and 1.0 μg Cd/L must be monitored for at least 30 minutes throughout three hours, as shown in **Figure 5**.

Subsequently, the normality and homogeneity of the data must be analyzed by applying the Shapiro–Wilk test with N-1 degrees of freedom and Levene test with degrees of freedom 1 and 2, determined as K-1 and (k-N)-K (N = absorbance

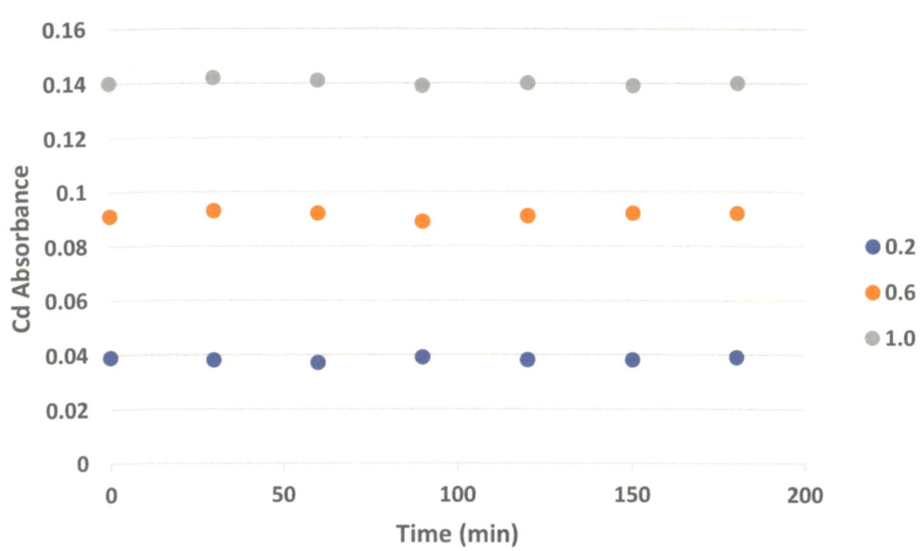

Figure 5.
Cadmium hollow cathode lamp stability [70].

readings and K = number of times). One-way ANOVA statistical treatment is applied next to determine whether there are any significant differences between the absorbances obtained for each standard during the three hours of analysis.

In the Shapiro–Wilk test, the Ho was accepted, since the calculated W was lower than the tabulated W (0.999) and the homogeneity of variances with calculated F was lower than the tabulated F (5.100), indicating that the means obtained are representative values of the absorbances of each one of the standards, with good variation coefficients (<3.5%) [70].

The one-way ANOVA used to measure the absorbance of each of the standards for the metal, showed a constant value for the sum of squares, as for the root mean square (0.000). We can also see that the calculated F is lower than the tabulated F at a 95% confidence level, indicating that there were no significant differences between the values of the absorbances at different times when the analysis was performed [70].

3.2 Optimizing the calcination and atomization temperature to determine cadmium by graphite furnace

In order to obtain a higher sensitivity for cadmium determination, the calcination and atomization temperature in the graphite furnace must be optimized. To do so, a standard of 1 μg/L of cadmium must be prepared and the atomization temperature set by varying the calcination temperature from 500 to 800 °C in 50 °C intervals. Once this is done, the highest absorbance is observed (which in our study was between 500 and 550 °C) and another variation of the calcination temperature in this range is made, but this time by modifying it in 10 °C intervals. Once the calcination temperature has been optimized, a standard of 0.8 μg/L of Cd is prepared and read on the atomic absorption equipment, setting the calcination temperature at 800 °C and varying the atomization temperature from 900 to 1800 °C. This variation is made at intervals of 100 °C. **Table 2** shows the optimum temperatures at 530 and 1750 °C [54, 70].

3.3 Quantification method

Cd quantitative analysis was performed using the calibration curve for which standard solutions of the metal were prepared as follows:

The calibration curve was constructed based on the stock solution of 1000 μg Cd/L by preparing 5 mL of six cadmium standards in a range of 0.1 to 1.0 μg Cd/L in 0.2% HNO3 solution. For each Cd standard, four absorbance readings are taken in the AAS-HG.

The Pearson's Correlation Coefficient showed that the calibration curve for cadmium has a high r value, above 0.9950, demonstrating that there is a positive correlation between the absorbance of the metal and its concentration [54, 70].

Stage	Temperature(°C)	Time(s)	Argon Gas Flow (L/min)
Drying	100	30	0.2
Calcination	530	20	0.2
Atomization	1750	3	Off
Cleaning	2500	3	0.2

Table 2.
Optimal graphite furnace temperature programming for Cd determination.

3.4 Standardization of the analytical method for cadmium determination

Statistical quality parameters must be evaluated in order to standardize the analytical method, as described below.

3.4.1 Linearity

To examine linearity, a calibration curve needs to be prepared in a concentration range including at least six different cadmium concentrations between 0.1 and 1.0 μg Cd/L. This should be analyzed using the atomic absorption equipment, taking four absorbance readings for each cadmium standard. The calibration curve should be plotted (Absorbance vs. concentration) and the correlation coefficient and slope should be statistically evaluated, through statistical treatment of the parametric data obtained for each concentration level [54, 70].

A regression analysis for the model $Y = \beta 0 + \beta 1X$ is required to confirm whether the degree of relationship is significantly linear between the two variables (absorbance and concentration). **Table 3** illustrates the coefficient that defines the slope of the regression line and the t-statistics for cadmium, making it possible to contrast the null hypothesis (Ho) that the slope has a value of zero. According to the results, the calculated statistics are greater than the tabulated ones, confirming that the slopes differ significantly from zero and therefore, the absorbance is significantly correlated to the concentration [54, 70].

The one-way ANOVA was applied to confirm the association between the two variables (absorbance and concentration) by means of a linear regression for cadmium. The results showed that the calculated F (6119.282) was higher than the tabulated F (4.080) and therefore, the null hypothesis raised is rejected (Ho = There is no significant linear relationship between absorbance and concentration) [54, 70].

3.4.2 Precision

Standard deviation (s) and the coefficient of variation (CV) must be determined in order to analyze the degree of agreement between the individual data obtained when the method is repeatedly applied to multiple aliquots of a homogeneous sample.

Accuracy is evaluated at two levels: Repeatability and Intermediate accuracy as described below.

3.4.2.1 Repeatability

We analyzed the precision obtained after performing five calibration curves, in a concentration range of 0.1 to 1.0 μg/L for cadmium, where absorbance is measured four times per standard, under the same operating conditions in a short time interval (same day) by the same analyst and using the same equipment, materials, and reagents. This value corresponds to 0.136 [54, 70].

Metal	Slope	Calculated T	Tabulated T
Cadmium	0.1324	78.226	2.048

Table 3.
Regression analysis for model $Y = \beta o + \beta 1x$.

3.4.2.2 Intermediate precision

We analyzed the precision obtained after performing seven calibration curves, in the same concentration range as for repeatability, where absorbance is measured four times per standard under the same operating conditions, in different time intervals (7 different days), by the same analyst and using the same equipment, materials, and reagents. This value corresponds to 0.135 [54, 70].

This method increases sensitivity when the analysis is conducted on the same day, indicating that the proposed method for sample preparation is appropriate. The sensitivity measured for the method was 0.136.

For the two previous procedures (Repeatability and intermediate precision), the Shapiro–Wilk test was applied with N-1 degrees of freedom (N = readings of absorbance) and Homogeneity of variances with degrees of freedom 1 and 2 calculated as K-1 and (k-N)-K respectively (where N = readings of absorbance and K = number of calibration curves), posing the corresponding null and alternative hypotheses at a 95% confidence level. In addition, for each of the (standard) concentration levels, the respective standard deviations and variation coefficients were obtained to determine whether the method is accurate.

For repeatability, the results show that at a 95% confidence level, the calculated W is lower than the tabulated W (0.999) for all levels of metal concentration, with 3 degrees of freedom for cadmium. Thus, the null hypothesis is accepted and it is concluded that the data come from a normal distribution. Applying Levene's test for cadmium data, we can see that for 4 and 15 degrees of freedom, the calculated statistic is lower than the tabulated statistic for all levels of concentration. Thus, we conclude that the variances are homogeneous. The standards present a low standard deviation (less than 0.004) and a coefficient of variation that is lower or equal to 5%, thus indicating that the method used presents a good repeatability [54, 70].

For the intermediate precision, the calculated Shapiro–Wilk statistic was lower than the tabulated statistic (0.999), therefore concluding that the data come from a normal distribution at a 95% confidence level. Taking into account Levene's statistic, the calculated statistic was lower than the tabulated one, thus concluding that the variances are homogeneous [54, 70].

The metal standards, show a small deviation (less than 0.005) and a coefficient of variation that is lower than or equal to 4.0%, leading to the conclusion that the method is of good accuracy [54, 70].

3.4.3 Sensitivity

Sensitivity is assessed as analytical sensitivity and calibration sensitivity, as described below.

3.4.3.1 Analytical sensitivity

The parameters used to determine analytical sensitivity are the limit of detection and quantification. Fifteen absorbance readings were taken from the metal target and the standard deviation calculated along with the detection and quantification limits, following the method suggested by IUPAC (1995) [70, 71].

The LOD of Cd was 0.02 mg/L (0.4 mg/kg) and corresponds to the minimum amount of Cd derived from the lowest analytical signal that can be detected with reasonable certainty. The LOQ of Cd was 0.07 mg/L (1.4 mg/kg) and represents the minimum concentration that can be measured with precision and accuracy. The LOD and LOQ are adequate for the quality control of biopolymers [54, 70].

3.4.3.2 Calibration sensitivity

This parameter is determined by comparing the slopes of the Cd calibration curves obtained with the precision of the method (repeatability and intermediate accuracy). The slopes obtained for repeatability (0.1358 ± 0.004) and intermediate precision (0.1351 ± 0.004), show that there is a greater response to concentration changes in repeatability [70].

3.4.4 Accuracy assessed as percent recovery

The accuracy of the method is determined in terms of percent recovery by adding in triplicate, known amounts (4, 8 and 12 μL) of a 1000 μg/L standard of cadmium to a thermoformed or flexible film sample prior to the digestion process. Following this, the respective readings (4 absorbance readings for cadmium) are taken using the atomic absorption equipment of the doped sample and of the sample with the standard added.

To calculate the percentage recovery, the cadmium concentrations in the samples are determined previously, as indicated in Eq. (2).

$$Recovery(\%) = \frac{Concentration\,with\,stock\,added - Concentration\,with\,no\,stock\,added}{Concentration\,with\,stock\,added} * 100 \quad (2)$$

The percent recovery of Cd ranged between 96.23% and 97.31% for the MPER 183 thermoformed material, which makes it possible to conclude that the extraction method used is suitable, and that the determinations are therefore reliable [54, 70].

A statistical t-student test should be applied to determine whether there are significant differences between the value obtained from recovery and the 100% level. The experimental t-values are compared with the tabulated values ($t(0.05, 8) = 1.860$) for 8 degrees of freedom in our case. The results show that for all the recoveries of the thermoformed samples, there are significant differences between the average values and 100% since t calculated > t tabulated (11.489> 1.860) [54, 70].

3.5 Stability of the metals

To establish the testing time for a standard and a sample in a laboratory, the stability of the samples over time must be determined as follows.

3.5.1 Standard stability

Cadmium standards are prepared (0.8 μg/L) and kept refrigerated while they are analyzed for ten consecutive days. A one-way ANOVA analysis (calculated F < tabulated F, 1,675 < 2,420) indicates that there are no significant differences between the average absorbances during the days analyzed, and therefore this standard can be prepared and stored at a temperature of 4 °C for analysis for up to 10 days [54, 70].

3.5.2 Stability of metals in digested samples

To evaluate the stability of cadmium, a thermoformed sample is subjected to digestion via a dilution of 5 to 100 mL, which is refrigerated and read on the equipment for ten consecutive days. The samples for flexible film were not read, as in our case, they were not detectable.

By applying a Shapiro–Wilk statistical analysis, we were able to conclude that the data come from a normal distribution, since the calculated W is smaller than the tabulated W and the data maintain a homogeneous distribution, since the calculated Levene statistic is smaller than the tabulated one [70].

A one-way ANOVA was applied to determine any significant differences in the average absorbances between the different days analyzed. The results show that cadmium was stable in the thermoformed sample analyzed in our study (CM 4574–7) as the calculated F < tabulated F (1.87 < 2,420) [70].

According to the results obtained, cadmium is stable in the thermoformed digested sample. Thus, these samples can be stored at 4 °C and analyzed on consecutive days [70], allowing the laboratory to establish its analysis times and organize the relevant protocols.

3.6 Determination of cadmium by atomic absorption spectrometry

3.6.1 Thermoformed and flexible films

To quantify cadmium in the samples, the graphite furnace atomic absorption spectrometer was used under the instrumental conditions shown in **Table 4**. The metal was measured in triplicate in the thermoformed and flexible film samples, as well as in the source materials (cassava flour, cassava starch and fique fiber), in order to determine whether the metal in the samples derives from the raw material used for manufacture.

Cd concentrations in the thermoformed products typically ranged between 4.2 mg/kg and 17.9 mg/kg, which could be the result of the quality of the raw materials used to process the biopolymers. Sample concentrations were significantly different from each other, given that when applying one-way ANOVA, the calculated F value was greater than tabulated F (1507.861 >3.501) at 95% confidence.

Cd was not detected in flexible films, meaning that no contamination was present.

The cadmium concentrations found in the thermoformed products were lower than those established by NTC-4096[1] (1 mg Cd/kg). Thus, these biopolymers can be used for these purposes, but care needs to be taken as this metal can bioaccumulate, causing serious environmental problems in the long term.

Method	Graphite Furnace-AAS
Atomization	Electrothermal
Element	Cd
Lamp	Hollow cathode Cd
Background correction	D2 Quadline
Wave length (nm)	228.8
Injection volume	20
Slit (nm)	0.5

Table 4.
Instrumental conditions in the GF-AAS for cadmium determination.

[1] The standard that regulates the maximum level of heavy metals in samples that are going to be in contact with food.

3.6.2 Raw material

Once the Cd concentration in the samples of thermoformed products had been determined, we proceeded to analyze the main raw material —cassava flour and fique fiber— in order to identify whether the metal found in the thermoformed products originated in the processing materials.

The results show that the flours presented Cd concentrations of between 3.5–18.2 mg/kg, while Cd concentration in the fique fiber was 7.2 mg/kg, meaning that it contributes the most Cd to the thermoformed film.

Given that the fique fiber and flour are agricultural products, the presence of cadmium in these samples may be due to the fact that this metal is used in herbicides used to control weeds such as Linuron or Dinuron. It can also be present in soil from municipal waste and the incineration of plastic materials, in fertilizers (phosphorous and nitrogenous fertilizers), pesticides and fungicides such as copper oxychloride and carbofurans among others, used to cultivate cassava [4, 10, 42, 72].

4. Conclusions

The GF-AAS method developed was efficient (highly sensitive and acceptable in terms of accuracy and reliability) to quantify Cd in thermoformed and biodegradable flexible films and cassava flour samples. The method is therefore reliable, with low variation coefficients, and limits of detection and quantification that indicate that the standardized method is optimal.

The concentrations that were found in the samples of thermoformed and flexible films, are below the amounts allowed for products that come into contact with food (1 mg/Kg), but, as mentioned above care needs to be taken as Cd bioaccumulation can lead to grave environmental problems.

Cd content found in thermoformed films is associated with prior contamination of the raw material (during cultivation, pretreatment, and/or transportation) or contamination derived from old machinery used for manufacturing.

This study also provides new data for food safety authorities and which broaden the existing knowledge of the contribution of raw materials in terms of Cd concentrations in biopolymers.

There is great potential in using biopolymers in packaging and food conservation, in terms of the value these materials add to agricultural activity and in helping to reduce nonbiodegradable plastics in the environment.

Acknowledgements

The authors would like to acknowledge Universidad del Cauca (Industrial Analysis Unit, BICAMSA and QPN laboratories) and the Ministry of Agriculture and Rural Development.

Conflict of interest

The authors declare no conflict of interest.

Author details

Maite Rada-Mendoza*, José Luis Arciniegas Herrera and Patricia Vélez Varela
Universidad del Cauca, Popayán, Colombia

*Address all correspondence to: mrada@unicauca.edu.co

References

[1] Vílchez-Vargas R. Eliminación de metales pesados de aguas subterráneas mediante sistemas de lechos sumergidos: Estudio microbiológico de las biopelículas [PhD thesis]. Spain: University of Granada; 2005.

[2] Núñez A, Martínez S, Moreno S, Cárdenas M, García G, Hernández J, Rodríguez A, Castillo I. Determinación de metales pesados (aluminio, plomo, cadmio y níquel) en rábano (*Raphanus sativus* L.), brócoli (*Brassica oleracea* L. var. italica) y calabacín (*Cucurbita pepo* L. var. italica). Universidad Autónoma de Nuevo León (UANL), México; 2008. p. 1-5.

[3] Graeme KA and Pollack CV. Heavy metal toxicity, part II: lead and metal fume fever. The Journal of emergency medicine. 1998; 16:2: 171-177. DOI: 10.1016/S0736-4679(97)00283-7

[4] Jarup L. Hazards of heavy metal contamination. British Medical Bulletin. 2003; 68:1: 167-182. DOI: 10.1093/bmb/ldg032.

[5] Gallego SM, Pena LB, Barcia RA, et al. Unravelling cadmium toxicity and tolerance in plants: insight into regulatory mechanisms. Environmental and Experimental Botany. 2012; 83: 33-46. DOI: 0.1016/j.envexpbot.2012.04.006

[6] Rodrigues A and Nascentes CC. Development of a simple method for the determination of lead in lipstick using alcaline solubilization and grafite furnace atomic absorption spectrometry. Talanta. 2013; 105: 272-277. DOI: 10.1016/j.talanta.2012.09.021

[7] Akter N, Khan RA, Tuhin MO, et al. Thermomechanical, barrier, and morphological properties of chitosan-reinforced starch-based biodegradable composite films. Journal of Thermoplastic Composite Materials. 2014; 27:7: 933-948. DOI: 10.1177/0892705712461512

[8] Attaran SA, Hassan A and Wahit MU. Materials for food packaging applications based on biobased polymer nanocomposites: a review. Journal of Thermoplastic Composite Materials. 2015; 30:2: 143-173. DOI: 10.1177/0892705715588801

[9] Weber CJ. Biobased packaging materials for the food industry. Dinamarca: The Royal Veterinary and Agricultural University. 2000. p. 11-136.

[10] LeCoultre DA. Meta-analysis and risk assessment of heavy metal uptake in common graden vegetables [Master thesis]. USA: East Tennessee State University; 2001.

[11] Bakkali K., Ramos N., Souhail B., Ballesteros E. Characterization of trace metals in vegetables by graphite furnace atomic absorption spectrometry after closed vessel microwave digestion. Food Chemistry. 2009; 116: 590-594. DOI: 10.1016/j.foodchem.2009.03.010

[12] Wambua P., Ivens J, Verpoest I. Natural fibres: can they replace glass in fibre reinforced plastics. Composites Science and Technology. 2003; 63: 1259-1264. DOI: 10.1016/S0266-3538(03)00096-4.

[13] Funke U, Bergthaller W, Lindhauer MG. Processing and characterization of biodegradable products based on starch. Polymer Degradation and Stability. 1998; 59:1-3: 293-296.

[14] Morales R, Candal M. Diseño y fabricación de un molde de termoformado utilizando herramientas CAD/CAE. Revista de la Facultad de Ingeniería Universidad Central de Venezuela. ISSN 0798-4065 vol.2. 2006.

[15] Merchán J., Ballesteros D, Jiménez I, Medina J, Álvarez O. Estudio de la biodegradación aerobia de almidón termoplástico (TPS). Suplemento de la Revista Latinoamericana de Metalurgia y Materiales. 2009; S1: 1: 39-44.

[16] Martins I, Magina S, Oliveira L, Freire C, Silvestre A, Neto C, Gandini A. New biocomposites based on thermoplastic starch and bacterial cellulose. Composites Science and Technology. 2009; 69: 2163-2168.

[17] Bilk AP, Ruffo S, Eiras MV, et al. Efficacy of some biodegradable films as pre-harvest covering material for guava. Scientia Horticulturae. 2011; 130:1: 341-343.

[18] Vicentino SL, Floriano PA, Dragunski DC, et al. Films of starch cassava to coat and conservation of grapes. Quimica Nova. 2011; 34:8: 1309-1314. DOI: 10.1590/ S0100-40422011000800003.

[19] Machado BAS, Nunes IL, Pereira FV, et al. Development and evaluation of the effectiveness of biodegradable films of cassava starch with nanocellulose as reinforcement and yerba mate extract as an additive antioxidant. Ciencia Rural. 2012; 42:11: 2085-2091. DOI: 10.1590/ S0103-84782012001100028.

[20] Montoya U, Zuluaga R, Castro C, et al. Development of composite films based on thermoplastic starch and cellulose microfibrils from colombian agroindustrial wastes. Journal of Thermoplastic Composite Materials. 2012; 27:3: 413-426. DOI: 0.1177/0892705712461663.

[21] Villada Castillo HS, Navia DP and Castañeda JP. Biodegradable films obtained from cassava starch and their manufacture process. Patent WO 2013042083 A1, Colombia, 2013.

[22] Costa SS, Druzian JI, Machado BAS, et al. Bi-functional biobased packing of the cassava starch, glycerol, licuri nanocellulose and red propolis. PLoS One. 2014; 9:11: e112554. 10.1371/ journal.pone.0112554.

[23] Santos RAL, Muller CMO, Grossmann MVE, et al. Starch/poly (butylene adipate-co-terephthalate)/ montmorillonite films produced by blow extrusion. Quimica Nova. 2014; 37:6:937-942. DOI: 10.5935/0100-4042.20140170

[24] Arrieta A, Jaramillo A and Palencia M. Conductive films from cassava starch as material for an electrochemical accumulator (battery). Revista de la Sociedad Química del Perú. 2015; 81:4: 328-338.

[25] Pagno CH, De Farias YB, Costa TMH, et al. Synthesis of biodegradable films with antioxidant properties based on cassava starch containing bixin nanocapsules. Journal of Food Science and Technology. 2016; 53:8: 3197-3205.

[26] Sueiro AC, Faria-Tischer PCS, Lonni AASG, et al. Biodegradable films of cassava starch, pullulan and bacterial celulose. Quimica Nova. 2016; 39:9: 1059-1064. DOI: 10.5935/0100-4042.20160118.

[27] Liu Y. Cassava starch degradable film, and preparation method thereof. Patent CN 106565998 A 20170419. 2017

[28] Medina-Jaramillo C, Ochoa-Yepes O, Bernal C, et al. Active and smart biodegradable packaging based on starch and natural extracts. Carbohydrate Polymers. 2017; 176: 187-194. DOI: 10.1016/j.carbpol.2017.08.079.

[29] Muller J, González-Martínez C and Chiralt A. Combination of poly(lactic) acid and starch for biodegradable food packaging. Materials. 2017; 10:8: E952. DOI: 10.3390/ma10080952.

[30] Dos Santos K, Lopes N, Haas T, et al. Characterization of active biodegradable films based on cassava starch and natural compounds. Food Packaging and Shelf Life. 2018; 16:138-147. DOI: 10.1016/j.fpsl.2018.03.006

[31] Ezeoha SL, Ezenwanne JN. Production of biodegradable plastic packaging film from cassava starch. IOSR Journal of Engineering. 2018; 3:10: 14-20. DOI:10.9790/3021-031051420

[32] Mann GS, Singh LP, Kumar P, et al. Green composites: a review of processing technologies and recent applications. Journal of Thermoplastic Composite Materials. 2018; 1-27. DOI: 10.1177/0892705718816354.

[33] Kaisangsri N, Kowalski R, Kerdchoechuen O, et al. Cellulose fiber enhances the physical characteristics of extruded biodegradable cassava starch foams. Industrial Crops and Products. 2019; 142:111810. DOI: 10.1016/j.indcrop.2019.111810

[34] De Lima C, Crepaldi M, De Oliveira O, et al. Biodegradable films based on commercial κ-carrageenan and cassava starch to achieve low production costs. International Journal of Biological Macromolecules. 2020; 165:582-590. DOI: 10.1016/j.ijbiomac.2020.09.150

[35] Meite N, Koffi L, Tohoue M, Goure B, et al. Properties of hydric and biodegradability of cassava starch-based bioplastics reinforced with thermally modified kaolin. Carbohydrate Polymers. 2021; 254: 117732. DOI: 10.1016/j.carbpol.2020.117322

[36] Adebowale AA, Olatunde OO, Adegunwa MO, et al. Mechanical and sensorial characteristics of cassava and yam composite starch films. Journal of Food Processing and Preservation. 2014; 38:4: 1994-**1998.** DOI:10.1111/jfpp.12175.

[37] Shimao M. Biodegradation of plastics. Current Opinion in Biotechnology. 2001; 12:3: 242-247. DOI: 10.1016/s0958-1669(00)00206-8.

[38] Ruíz G, Montoya C, Paniagua M. Degradabilidad de un polímero de almidón de yuca. Revista EIA Escuela de Ingeniería de Antioquia. 2009; 12: 67-78.

[39] Velazquez G. Efecto de la humedad y la temperatura sobre Las propiedades mecánicas y de transporte de películas flexibles. Universidad Autónoma de Queretaro Facultad de Química. México. 2000.

[40] La Fuente C, Tamyris A, Tadini C, Duartes P. Ozonation of cassava starch to produce biodegradable films. International Journal of Biological Macromolecules. 2019; 141; 1:713-720. DOI: 10.1016/j.ijbiomac.2019.09.028

[41] Obanijesu EO and Olajide JO. Trace metal pollution study on cassava flour's roadside drying technique in Nigeria. In: Ek Yanful, editors. Appropriate technologies for environmental protection in the developing world. Dordrecht: Springer; 2009. p. 333-339.

[42] Idodo-Umeh G, Ogbeibu AE. Bioaccumulation of the heavy metals in cassava tubers and plantain fruits grown in soils impacted with petroleum and non-petroleum activities. Research Journal of Environmental Sciences. 2010; 4:1: 33-41. DOI: 10.3923/rjes.2010.33.41.

[43] Di Toppi LS and Gabbrielli R. Response to cadmium in higher plants. Environmental and Experimental Botany. 1999; 41:2: 105-130.

[44] ATSDR. Agency for Toxic Substance and Disease Registry. Toxicological profile for cadmium. Report, US Department of Health and Human Services, Public Health Service, Centers for Disease Control. Atlanta, USA, 2012.

[45] Thijssen S, Cuypers A, Maringwa J, et al. Low cadmium exposure triggers

a biphasic oxidative stress response in mice kidneys. Toxicology. 2007; 236:1-2: 29-41. DOI: 10.1016/j.tox.2007.03.022.

[46] Rizwan M, Ali S, Adrees M, et al. A critical review on effects, tolerance mechanisms and management of cadmium in vegetables. Chemosphere. 2017; 182: 90-105. DOI: 10.1016/j.chemosphere.2017.05.013

[47] John MK, VanLaerhoven CJ, Chuah HH. Factors affecting plant uptake and phytotoxicity of cadmium added to soils. Environmental Science & Technology. 1972; 6:12: 1005-1009. DOI: 10.1021/es60071a008

[48] Anwar S, Nawaz MF, Gul S, et al. Uptake and distribution of minerals and heavy metals in commonly grown leafy vegetable species irrigated with sewage water. Environmental Monitoring and Assessment. 2016; 188:9: 541. DOI: 10.1007/s10661-016-5560-4

[49] Ahmad JU and Goni MA. Heavy metal contamination in water, soil, and vegetables of the industrial areas in Dhaka, Bangladesh. Environmental Monitoring and Assessment. 2010; 166:1-4: 347-357. DOI: 10.1007/s10661-009-1006-6.

[50] Kim HK, Jang TI, Kim SM. Impact of domestic wastewater irrigation on heavy metal contamination in soil and vegetables. Environmental Earth Sciences. 2015; 73:5: 2377-2383. DOI:10.1007/s12665-014-3581-2

[51] Antoine JMR, Fung LAH and Grant CN. Assessment of the potential health risks associated with the aluminium, arsenic, cadmium and lead content in selected fruits and vegetables grown in Jamaica. Toxicology Reports. 2017; 4: 181-187. DOI: 10.1016/j.toxrep.2017.03.006

[52] Ueno D, Yamaji N, Kono I, et al. Gene limiting cadmium accumulation in rice. Proceedings of the National Academy of Sciences of the United States of America. 2010; 107:38: 16500-16505. DOI: 10.1073/pnas.1005396107.

[53] ICONTEC. Norma Técnica Colombiana. 2006. Plásticos. Plastificantes DOP y DOA grado alimento.

[54] Rada-Mendoza M, Arciniegas-Herrera JL, Hoyos-Saavedra OL, Del Castillo R, Chito-Trujillo D. Atomic absorption spectrometry for the quantification of cadmium in thermoformed and biodegradable flexible films made from cassava (Manihot esculenta crantz). Journal of Thermoplastic Composite Materials. 2019: 1-14.

[55] CHA J, Chung D, Seib P, Flores R, Hanna M. Physical properties of starch-based foams as affected by extrusion temperature and moisture content. Industrial Crops and Products. 2001; 14:1: 23-30. DOI: 10.1016/S0926-6690(00)00085-6

[56] Gunaratne A, Hoover R. Effect of heat moisture treatment on the structure and physicochemical properties of tuber and root starches. Carbohydrate Polymers. 2002; 49: 4: 425-437. DOI: 10.1016/S0144-8617(01)00354-X.

[57] Tabi T. The analysis of processing and usage of injection moulded biodegradable polymers made of starch and poly(lactic acid). [Thesis Doctoral]. Budapest: Budapest University of Technology and Economics, Departamento de Ingenieria de Polímeros; 2010. p. 3, 10

[58] AOAC. Association of official analytical chemists. Determination of moisture content. In: Official Methods of Analysis. AOAC 950.43. 18th ed. Gaithersburg, MD: The association, 2005.

[59] Sanches M, Vilanova E. Técnicas analíticas de contaminantes químicos. España: Diaz de Santos; 2004. p. 12-13.

[60] Valcárcel, M. Técnicas analíticas de separación. Barcelona: Reverté. 1988. p. 164-165

[61] Gutiérrez H, Salazar R. Análisis y Diseño de Experimentos. México: Mc Graw-Hill; 2004. p.103-105

[62] Marin B, Chopin E, Jupinet B., Gauthier, D. Comparison of microwave-assisted digestion procedures for total trace element content determination in calcareous soils. Talanta. 2008; 77: 282-288. DOI: 10.1016/j.talanta.2008.06.023.

[63] MARK, J. Polymer Data Handbook. Oxford University Press, ed. 1, 1999, p. 145, 707

[64] CROMPTO R. Determination of Additives in Polymers and Rubbers. Rapra Technolog, ed. 1, 2007, p. 337-345.

[65] Alvaro T., Morgana B. Marcia M., Goreti R. and Welz B. Determination of cadmium and lead in plastic material from waste electronic equipment using solid sampling graphite furnace atomic absorption spectrometry. Microchemical Journal. 2010; 96: 102-107. DOI: 10.1016/j.microc.2010.02.008

[66] Roca de Togores M, Farré R and Frigola AM. Cadmium and lead in infant cereals- electrothermal-atomic absorption spectroscopic determination. Science of the Total Environment. 1999; 234:1-3: 197-201. DOI: 10.1016/s0048-9697(99)00260-0.

[67] Bentlin F, Pozebon D, Mello P, et al. Determination of trace elements in paints by direct sampling graphite furnace atomic absorption spectrometry. Analytica Chimica Acta. 2007; 602:1: 23-31. DOI: 10.1016/j.aca.2007.09.003

[68] Sudunagunta D, Venkatesh N and Meyyanathan D. Atomic absorption spectroscopy: a special emphasis on pharmaceutical and other applications. Journal of Pharmacy Research. 2012; 5:3: 1614-1619.

[69] Junior MMS, Silva LOB, Leao DJ, et al. Analytical strategies for determination of cadmium in brazilian vinegar samples using ET AAS. Food Chemistry. 2014; 160: 209-213. DOI: 10.1016/j.foodchem.2014.03.090.

[70] Del Castillo and Rada-Mendoza M. Determinación de mercurio y cadmio en empaques termoformados y películas flexibles biodegradables, empleando la técnica espectrométrica de absorción atómica [thesis]. Popayán: Universidad del Cauca; 2012.

[71] Sardans J, Montes F, Peñuelas J. Determination of As, Cd, Cu, Hg and Pb in biological samples by modern electrothermal atomic absorption spectrometry. Spectrochimica Acta Part B Atomic Spectroscopy. 2010; 65: 97-112. DOI: 10.1016/j.sab.2009.11.009.

[72] Santos MC, Nobrega J, Baccan N, Cadore S. Determination of toxic elements in plastics from waste electrical and electronic equipment by slurry sampling electrothermal atomic absorption spectrometry. Talanta. 2010; 81:1781-1787. DOI:10.1016/j.talanta.2010.03.038

Chapter 4

Simultaneously Recovery of Phosphorus and Potassium Using Bubble Column Reactor as Struvite-K and Implementation on Crop Growth

Endar Hidayat and Hiroyuki Harada

Abstract

Struvite-K, similar to NH_4-struvite with a composition of Mg:K:P (1:1:1). It is called struvite-K because the K replaces the NH_4 in struvite. The composition usually used as fertilizer and can be recycling from wastewater including livestock wastewaters. In addition, Struvite-K which tends to form scale on surfaces of equipment which problem in many industries. The present study was used bubble column reactor which simple and efficient. In addition, the process can be implementation in wastewater industry which low-tech processes. Then, the struvite-K precipitate was implementation on crop growth which compared with coffee husk compost. The results show the removal of P via struvite-K showed 98.5% with the precipitation Mg:P of 0.7 and K:P of 1 with yields of 11.28 gram. Increases of magnesium dosage which decreases of P removal rate and affected of crystal size structure. Compost and struvite-K have similar positive impact on crop growth of (radish and komatsuna) were compared than control. In the other hand, the struvite-K is more effective than compost. This might be indicated that struvite-K is more slow-release nutrient than compost and higher macro nutrient supplied on soil which crop needed.

Keywords: Struvite-K, precipitates, soil, recovery of wastewater, fertilizers

1. Introduction

1.1 Bubble column reactor for struvite-K precipitation

Magnesium potassium phosphate hexahydrate, commonly known as struvite-K, is a sparingly crystallize consisting of equal molar amount of magnesium potassium and phosphate and six water of hydration, hence the chemical formula is $MgKPO_4.6H_2O$. Struvite-K is formed according to the following reaction:

$$Mg^{2+} + K^{+} + PO_4^{\ 3-} \rightarrow MgKPO_4.6H_2O \tag{1}$$

Struvite-K has been synthesized to show its viability for phosphorus and potassium removal from wastewater [1–4] and naturally occurring. Tanaka [5] reported that phosphorus and potassium from effluents of livestock wastewater contains (5.5 mM) and (63.9 mM), respectively. This is a new resource for potassium and phosphorus demand for use as fertilizers which following global population in every year. It was estimated the world population to reach between 9 and 10 billion by 2050 [6, 7].

Struvite-K precipitates as a white powder needle-like structure, which tends to form scale on surfaces of equipment. Struvite-K scaling is a well-recognized problem in many industrial processes and domestic applications. The problem is that not only from pipes but also pumps, centrifuges and aerators can be blocked by struvite [8–10]. In the other hand, this is not problem but solution to supplying nutrient on crop growth.

Struvite-K, similar to NH_4-struvite has a desired low solubility and chemical composition of Mg:K:P (1:1:1). It also has a high market value in the agricultural area (Sean 2018). It is called struvite-K because the K replaces the NH_4 in struvite ($MgNH_4PO_4.6H_2O$). The present study, we used synthetic wastewater was designed using bubble column reactor to simulate effluent from livestock wastewater to recovery of phosphorus and potassium as struvite-K which used as fertilizers for increasing global food production.

1.2 Implementation on crop growth

Most of the nutrients absorbed by plants come from organic matter. Therefore, the unique fertilizer formula comes from compost or struvite-K precipitate. They provide a rich source of nutrients and can be added to the soil to promote the growth of various crops. Therefore, compost and struvite-K precipitation are intended to serve the society by increasing farmers' incomes to revitalize the soil and increase farm yields. We fertilized the two fertilizers from compost and struvite-K precipitate, in order to evaluate the effects of fertilizer on soil and yields of radish (*Raphanus sativus* L.) and Komatsuna (*Brassica rapa var. perviridis*).

*1.2.1 Radish (*Raphanus sativus *L.)*

Radish (*Raphanus sativus L.*) belongs to the genus and species of *Radish* in the *cruciferous* family. Radishes are grown and consumed all over the world and are considered part of the human diet, although it is not common in certain populations. It is one of the most important and popular root vegetables in tropical, subtropical and temperate regions of the world (including Japan). It is grown as an annual and biannual vegetable crop, depending on its planting purpose. Radishes are mainly cold-season vegetable crops. Asian varieties can tolerate higher temperatures than European varieties. In a mild climate, radishes can be grown almost all year except for the summer months [10]. Its young roots can be eaten raw in salads or cooked as vegetables. It has a spicy taste and is considered an appetizer. Young leaves can also be cooked and eaten as vegetables. Radish preparations are useful for liver and gallbladder diseases. Roots, leaves, flowers and pods are active against Gram-positive bacteria, urinary system diseases, hemorrhoids and stomach pain. In addition, the salt extracted from the roots can be dried and burned into white ash, which can be used to relieve stomach problems [11].

1.2.2 Komatsuna (Japanese mustard spinach)

Japanese mustard spinach (*Brassica rapa var. perviridis*) is also called Komatsuna in Japanese. It is a common and popular leafy vegetable in the Japanese diet.

Komatsuna contains low energy and high nutrients, which is very effective in lowering serum cholesterol. On the other hand, Komatsuna has a compound called sulforaphane, which can help our body fight cancer. Sulforaphane actively kills cancer stem cells and slows the growth of tumors [12]. Komatsu greens are rich in calcium and are commonly used in kimchi in Japan and as feed crops in many Asian countries.

In 2019, the production of Komatsuna in Japan increased by 26.5% compared with 2006, and the planting area increased by 29.2%, with a total output of 114,900 tons [13]. Komatsuna can be worn in relatively temperate regions most of the year, but it is usually grown as a cool seasonal crop (spring and autumn). It can tolerate some extreme cold and hot conditions, but it cannot stand for long periods of time [14]. Economically important members of this family include vegetables such as broccoli, cabbage, chinese cabbage, radish, cauliflower as well as the oil crop canola [15].

2. Materials and methods

2.1 Materials

Sodium dihydrogen phosphate, magnesium chloride and sodium hydroxide were obtained from the Kanto Chemical Co. Inc. (Tokyo/Japan), potassium chloride was obtained from Wako Pure Chemical Industries Ltd. All chemicals and reagents were of analytical grade and used without further purification.

The soil was collected at 0–20 cm of depth in the field center of Prefectural University of Hiroshima. The soil was analyzed in the laboratory at the Department of Environmental Science, Prefectural University of Hiroshima, Japan. Elemental characteristics of soil and compost were described previous work [16]. Radish (*Raphanus sativus L.*) and komatsuna (*Brassica rapa var. perviridis*) were collected from market store, which is in the Shobara city, Hiroshima Prefecture.

2.2 Experimental design and treatments.

2.2.1 Bubble column reactor for struvite-K precipitation

The use of different magnesium additives to recover $MgKPO_4.6H_2O$ (MPP) was investigated. The experimental conditions are summarized in **Table 1**. Illustrates the experimental setup of the bubble column used in this study which has capacity of 10 L (**Figure 1**). There was a draught tube structure inside the bubble column. Air was fed using an air pump from the bottom of the tower as instead of mixing for homogeneous of solution. A pH probe was placed directly into the reactor just below the liquid surface. The bubble column was first filled with synthetic wastewater, and then potassium chloride, magnesium chloride and sodium dihydrogen phosphate solutions were fed from the outer reactor to the inner tubes. The pH

Run no.	Mg^{2+} mM	PO_4-P mM	K^+ mM	Mg/P initial
1	5.2	6.4	13.4	0.8
2	6.9	6.4	13.4	1.08
3	7.8	6.4	13.4	1.22

Table 1.
Characteristic of raw water.

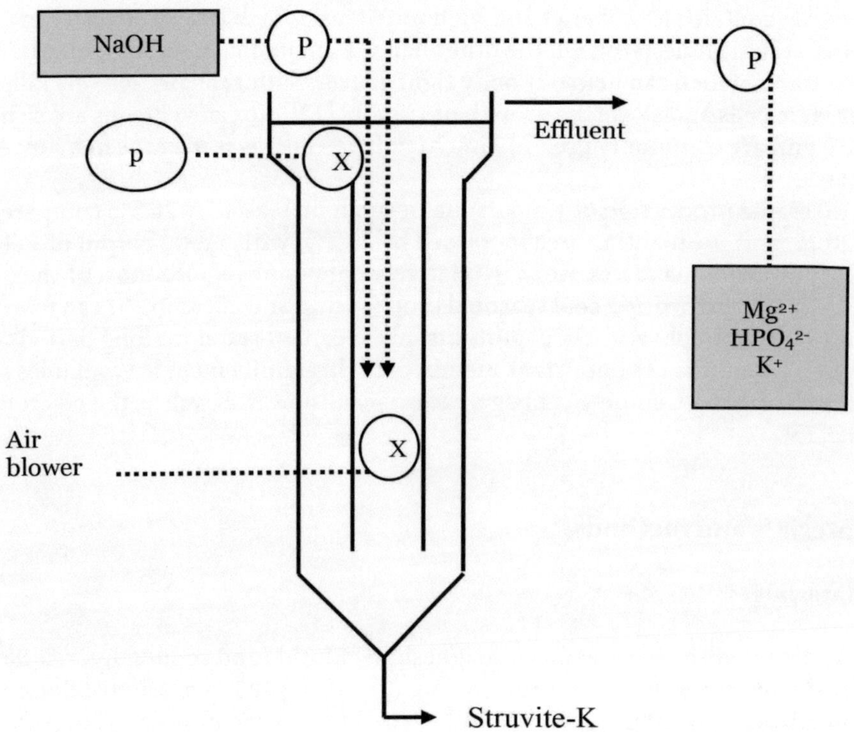

Figure 1.
Experimental set-up.

was adjusted to 12.9 with 0.1 M NaOH solutions. Afterward the solution of potassium chloride, magnesium chloride and sodium dihydrogen phosphate were added continuously with had pH of 3.4 by dose pump until pH of 11 with retention time of 1.98 (defined Eq. (2)).

$$HRT = \text{Total volume reactor}(L) / \text{Influent flow rate}(L) \quad (2)$$

Set points for minimum and maximum pH values defined a narrow band of 0.3 pH units. After filtering the reactor solution, a white precipitation was obtained; it was desiccated at 60°C for 24 h. Samples were taken directly from the precipitation zone. All experiments were done at room temperature.

2.2.2 Pot treatment designs

The experimental design was completely randomized design, with five treatments and three replications were presented by pots. Seeds were soaked in water for 24 h before sowing at a maximum depth of 1.2 cm. The equal proportions of compost samples (150 g and 10 seeds) i,e Radish (*Raphanus sativus L.*) and komatsuna (*Brassica rapa var. perviridis*) were filled and installed in a greenhouse. The LED model (PF15-S5WT8-D with power 5 W) was used as a light source with free space between lamps and a pot about 37 cm. Treatments consisted of C: 100% (control), A1: 0.1% of compost, A2: 0.3% of compost, B1: 0.1% of struvite-K and B2: 0.3% of struvite-K. The pots were watered periodically to prevent drought stress of the plants. The experimental were conducted for 9 days.

2.2.3 Analytical methods of struvite-K precipitates

The concentrations of PO_4^{3-} were determined by standard method (Japan Industrial Standard method JIS KO 102). The concentrations of potassium and magnesium ions were measured using atomic absorption spectrophotometer (AA-6300, Shimadzu Kyoto, Japan). The white precipitation was dissolved in 0.5 M HNO_3 in 1 h for determination of the crystalline components. Microscope images (Olympus CX-32) was used to observe the surface morphology of the crystals. The percent P removal and P recovery are in Eqs. (3) and (4), respectively.

$$P\,removal\% = \frac{P\,Initial - P\,equilibrium}{P\,Initial}100 \tag{3}$$

$$P\,recovery\% = \frac{P\,in\,white\,precipitation}{P\,decrement}100 \tag{4}$$

2.2.4 Analytical methods of soil

The sample was ground by using a coffee mill to pass through a sieve <2-mm before analysis [16]. Cation exchange capacity (CEC) were extracted 1 M NH_4OAc pH 7 via NH_4 with the procedure in below:

$$\text{CEC}\left(\text{meq}/100\,\text{g}\right) = 13.7 * \text{NH}_4\left(\text{mg}/\text{L}\right) * \text{dilution}/3 \tag{5}$$

The ammonium (NH_4) was measured by the phenate spectrophotometric with UV–Vis Spectrophotometer (Jasco V-530) at 640 nm. The detailed procedure in below:

Solution A

1. Phenol 5 gram
2. Sodium nitriferricyonide dehydrate 0.1 gram
3. Homogenized and adjusted with dilution water 500 mL

Solution B

1. 200 ml of NaCLO (assay 5%) = 200/5 = 40 mL
2. NaOH = 15 gram
3. Homogenized and adjusted with dilution water 1000 mL

Procedure:

10 mL from the liquid sample and 5 ml from solution A and B, respectively. Afterward, waiting for 30 minutes before analysis. Standard solution was used NH_4Cl with calibration curve, y = 0.1747x – 0.1465. R^2 = 0.9999.

3. Results and discussion

3.1 Effect different initial of Mg:P ratios on % P removal and precipitate

Bubble column reactor was used in laboratory studies with hopefully, it can be implementation in wastewater industry. From a structural point of view, they are very simple units equipped with added a dose pump as mixing system that allows for the homogenization of the wastewater with the reactants. The mixing condition inside the reactor represents a fundamental aspect because it affects the struvite formation [17]. An effective mixing promotes the crystals nucleation and growth by improving the mass of ions from the solution to the solid phase. Generally, completely mixed reactors can operate continuously or in batch mode. A batch column reactor works according to a series of phase and the struvite-K production and precipitation occur in the same unit (**Figure 2**).

In this study, we used $MgCl_2.6H_2O$ as magnesium source which have advantages of being very soluble allowing to recovery a precipitate with a high purity degree [18]. Due to easy management was used in many studies through which the kinetics of the nucleation process have been assessed [19, 20]. While the pH was maintained in alkaline condition of 11 which following based on [21]. Since the conditions favorable precipitation of struvite-K. Based on **Figure 3** shows that P removal rate was decreased from 98.5–80% which increasing of molar ratios of Mg:P from 0.8 to 1.22 since strong influence of precipitation in many scientist [22]. While for hydraulic retention time we maintained at 1.98 h; since in around of this was much more consistent than the rate reported in the literature and no effect with difference of HRT in precipitation [3, 4]. Furthermore, P recovery of struvite-K we conducted only for Mg:P of 0,8 since these concentrations was effective for % P removal than others with precipitate of Mg:P of 0,7 and K:P of 1 are shown in **Table 2**.

The optimal molar ratios must be assessed case-by-case as it depends heavily on the chemical–physical characteristics of wastewater and the reactor used. On the contrary, other works in which unconventional reagents were exploited, found that greater dosages are necessary. For example, Quintana et al. [23] found a strong influence a Mg:P molar ratio on the abatement of phosphate amount and the major removal was detected when used MgO dosed at molar ratio of 1.5. Moreover, the authors observed that the increase of the molar ratio promoted the removal rate growth (Quintana et al. [24]). In agreement with this consideration, Jaffer et al. [25] affirmed that with Mg:P addition lower than 1.05 the precipitation resulted was greater to avoid the formation of magnesium sodium phosphate [4].

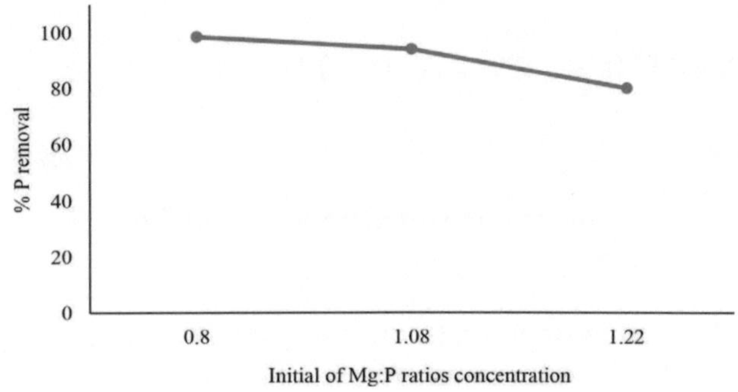

Figure 2.
Percentage of P removal with different of initial of Mg:P ratios concentration.

DOI: http://dx.doi.org/10.5772/intechopen.100103

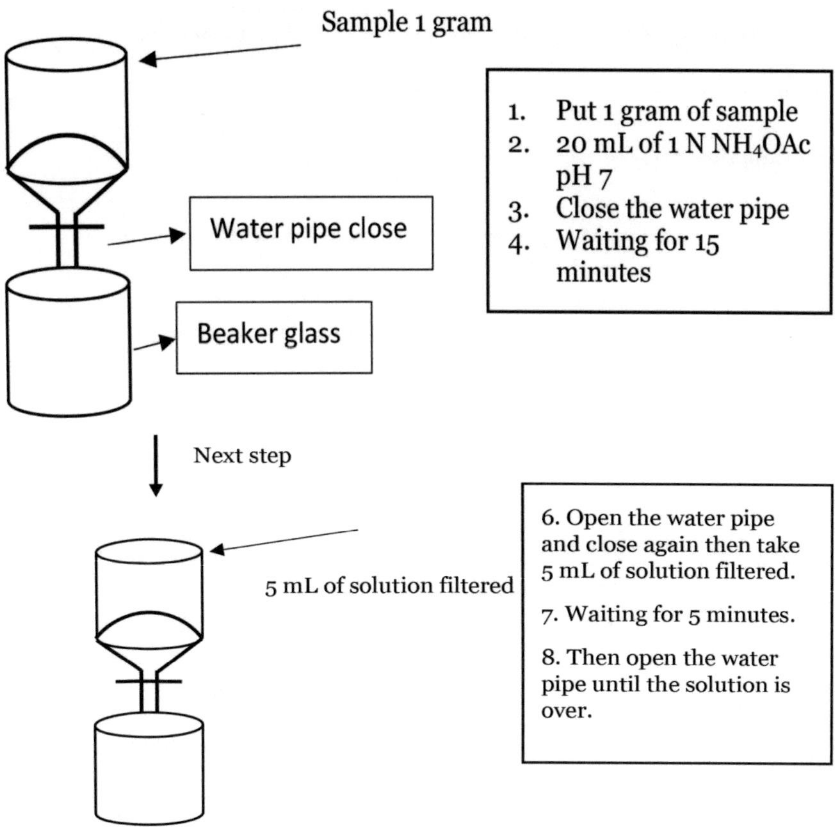

b. Extracted by using 10% NaCL
1. After filtered of solution is over, measure CEC via NH_4
2. Make a solution 10% NaCL

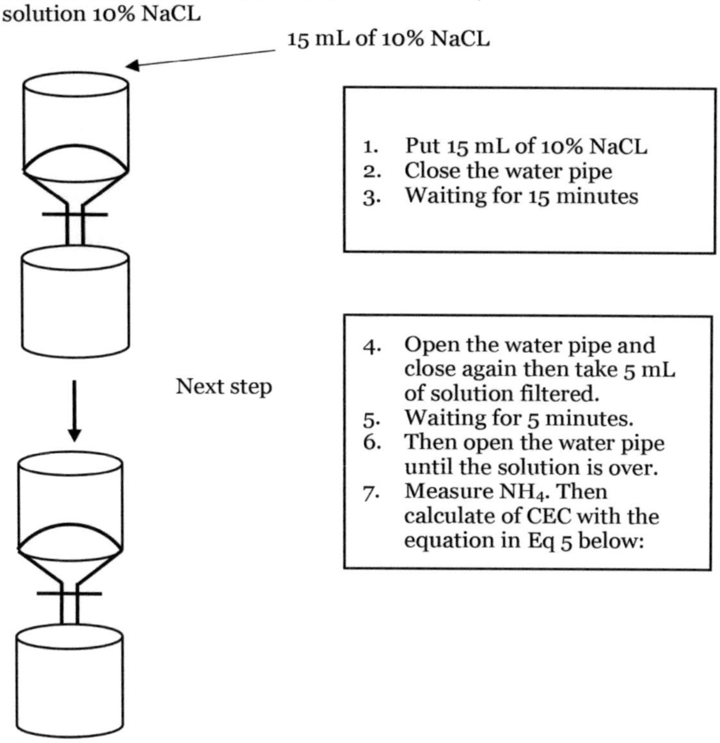

Figure 3.
Procedure for measure exchangeable cation and CEC via NH_4.

pH	Dosage Mg:P ratio	Mg:P (precipitate)	K:P (precipitate)	% P recovery
11	0.8	0.7	1	99.6

Table 2.
Effect of the magnesium dosage on the Mg:P molar concentration ratio in precipitate.

However, an excess of Mg can cause the formation of magnesium phosphate and reducing the precipitation of struvite which following for P removal rate. Korchef et al. [26] observed the phosphate removal caused by precipitation for Mg:P molar ratio under 4 and newberyite ($MgHPO_4.3H_2O$) and cattiite ($Mg_3(PO_4)_2.22H2O$) formation for Mg:P = 5.

3.2 Morphology of struvite-K precipitation

The morphology of struvite-K was observed by using Microscope images (Olympus CX-32). **Figure 4A** shows needle-like structure. This is agreement with some reported [3, 4, 27, 28]. While for **Figure 4B**, the needle-like was mixed with block and hexagonal structure. Furthermore, there are more blocks and hexagons seen in **Figure 4C**. This shows that as the amount of magnesium increases and the removal rate of phosphorus decreases, the structure is affected which mixed with another structure. The increase in the magnesium dosage also reduces the crystal size. The average of crystal size of 22.45 μm (**Figure 4A**). This is agreement with [29] that increase magnesium dosage reduced the average crystal size. Struvite-K forms a white crystalline solid. When dried in 60°C the struvite-K precipitated into appearance of white powder as shown in **Figure 4D**. While for the struvite-K yields of 11.28 gram.

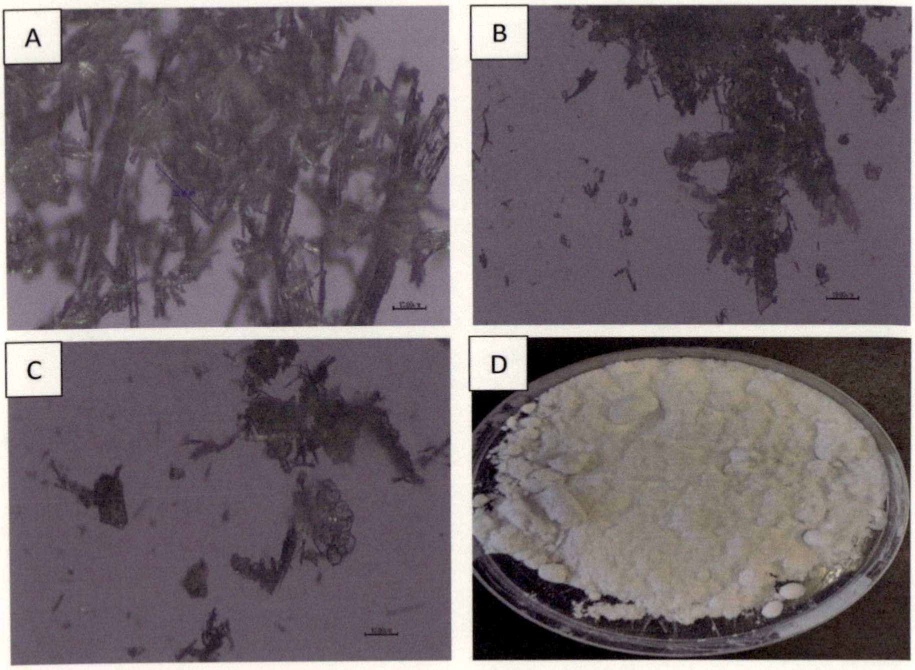

Figure 4.
Struvite-K crystals as seen x10/0.25 of Olympus. Image captured with microscope of Olympus CX-32. (A) Initial Mg:P of 0.8, (B) initial Mg:P 1.08, (C) initial Mg:P of 1.22 and (D) image of struvite-K precipitated dried (crystal of white powder).

3.3 Effect of fertilizers on soil cation exchange capacity

The cation exchange capacity (CEC) is a very important soil property for nutrient retention and supply and acts as a bridge between soil and plant [30]. **Table 3** presents the characteristics of soil CEC on two-kind of crop growth (radish and komatsuna). The results shows that all treatment of fertilizer (compost and struvite-K) with two-kind of crop growth were higher than control (soil only). However, the highest results seen in treatment of struvite-K with 0.3% percent ratio of soil for both of crop growth. This is indicated that struvite-K is effective as supplying of nutrient sources which have higher macro nutrient such as phosphate, magnesium and potassium. This is agreement with [31] who found application of phosphate fertilizer increasing of soil CEC after 40 years of application. This can be explained that phosphate, potassium and magnesium is the most important factor affecting on soil CEC.

3.4 Effect of fertilizers on yields of crop growth

As shown in **Table 4** presents the stem height, leaf area of radish. Overall, the treatment of fertilizers was highest than control (soil only).

In the treatment of compost, the highest rate of 0.3% with 4.46 and 0.98 cm on stem height and leaf area, respectively. While for treatment of struvite-K the

Treatment	CEC (meq/100 g)
Komatsuna	
Control	55.3
Compost 0.1%	71.8
Compost 0.3%	74.1
Struvite-K 0.1%	72.5
Struvite-K 0.3%	94.0
Radish	
Control	57.1
Compost 0.1%	67.8
Compost 0.3%	72.3
Struvite-K 0.1%	69.1
Struvite-K 0.3%	86.4

Table 3.
Characteristics of soil cation exchange capacity.

Treatment	Stem length (cm)	Leaf wide (cm)
Control	2.67	0.43
Compost 0.1%	3.08	0.68
Compost 0.3%	4.46	0.98
struvite-K 0.1%	4.38	0.8
struvite-K 0.3%	6.36	1.31

Table 4.
*Radish (*Raphanus sativus L.*).*

highest rate of 0.3% with 6.36 and 1.31 cm on stem height and leaf area, respectively. For images of radish on pot treatment as shown in **Figure 5**. Both of fertilizers were given benefits effect on yields of radish and positive correlation with soil CEC. If we comparison, the struvite-K was given more benefits effect which cause the struvite-K is higher contains of macro content such as magnesium, potassium and phosphate as nutrient sources on crop growth. Furthermore, **Table 5** presents the stem height and leaf area of komatsuna. The same dosage treatment with radish between 0.1% and 0.3%. Overall, the supplying of fertilizers was given benefits effect on yields of komatsuna. However, the more benefits in the treatment of

Figure 5.
Images of radish on pot treatment; (C) 100% (control), (A1) 0.1% of compost, (A2) 0.3% of compost, (B1) 0.1% of struvite-K and (B2) 0.3% of struvite-K.

Treatment	Stem length	Leaf wide
Control	3.43	0.58
Compost 0.1%	4.42	0.87
Compost 0.3%	5.4	1.22
struvite-K 0.1%	4.36	0.99
struvite-K 0.3%	5.74	1.27

Table 5.
Komatsuna (Japanese mustard plant).

Figure 6.
Images of komatsuna on pot treatment, (C) 100% (control), (A1) 0.1% of compost, (A2) 0.3% of compost, (B1) 0.1% of struvite-K and (B2) 0.3% of struvite-K.

struvite-K on dosage of 0.3% with 5.24 and 1.27 cm, on stem length and leaf wide, respectively. This might be indicated that struvite-K is more slow-release than compost and higher macro nutrient supplied on soil which needed on crop. Slow-release fertilizer which releases nutrients slowly over a long release time. Furthermore, it can reduce the number fertilization times and amount of fertilizer applied, which have good impact on fertilizer use efficiency [32–35]. Images of komatsuna on pot treatment as shown in **Figure 6**.

4. Conclusion

The present study used bubble column reactor which simple and efficient. Removal of P via struvite-K showed 98.5% with the precipitates Mg:P of 0.7 and K:P of 1. The yields of 11.28 gram. Compost and struvite-K have positive impact on crop growth of (radish and komatsuna) were compared than control. Which might be caused supplied nutrient source on soil and uptake on crop. However, the struvite-K is more effective than compost may cause contains higher macro nutrient such as magnesium, potassium and phosphate. This is indicated that the recovery process of P and K via struvite-K using bubble column reactor was very effective and efficient to utilization as a fertilizer.

Acknowledgements

The author (E.H) would like to thanks to Japanese Government (MEXT Scholarship) for financial support during studying in Prefectural University of Hiroshima.

Conflict of interest

The authors declare no conflict of interest.

Author details

Endar Hidayat* and Hiroyuki Harada
Department of Environmental Science, Prefectural University of Hiroshima, Shobara, Hiroshima, Japan

*Address all correspondence to: hidayatendar1@gmail.com

References

[1] Bennet, A. Potential for potassium recovery as K-Struvite. [thesis]. University of British Columbia; 2015.

[2] Zeng, Le, and Xiaomei Li. Nutrient removal from anaerobically digested cattle manure by struvite precipitation. Journal of Environmental Engineering Science. 2006; 5(4); 285-294.

[3] Wilsenach, JA, Schuurbiers CAH, van Loosdrecht MCM. Phosphate and potassium recovery from source separated urine through struvite precipitation. Water Research 2007;41(2): 458-466.

[4] Satoshi Y, Ohura S, Harada H, Akagi K, Yoshiharu M, Kawakita H, and Biplob KB. Simultaneous crystallization of phosphate and potassium as magnesium potassium phosphate using bubble column reactor with draught tube. Journal of Environmental Chemical Engineering: 2013;1(4):1154-1158. http://dx.doi.org/10.1016/j.jece.2013.08.032.

[5] Tanaka T, Koike N, Sto T, Arai T, Taira N. Recovery of phosphate from livestock wastewater by electrolysis. Journal of Japan Society On Water Environment. 2009;32(2): 79-85.

[6] Pew Research Center. Attitudes about aging: A global perspective [Internet]. 2014. Available from; http://www.pewglobal.org/2014/01/30/attitudes-about-aging-a-global-perspective/. [Accessed:2021-06-28]

[7] UN News Centre. World population projected to reach 9.6 billion by 2050 [Internet]. 2013. Available from; http://www.un.org/apps/news/story.asp?NewsID=45165#.VPd1E0LVmoU. [Accessed:2021-06-28]

[8] Parsons SA, Doyle JD. Struvite scale formation and control. Water Science & Technology. 2004; 49(2):177-1782. https://doi.org/10.2166/ wst.2004.0118

[9] Mamais D, Pitt PA, Yao WC, Loicano J, Jenkins D. Determination of ferric chloride dose to control struvite precipitation in anaerobic digesters. Water Environment Research. 1994:66:912-918. https://doi.org/10.2175/wer.66.7.8

[10] Umar UM, Ibrahim I, Iro, Obidola SM. Growth and yield of radish (*Raphanus sativus L.*) as influenced by different levels of kalli organic fertilizer on the Jos Plateau. Asian Journal of Research in Crop Science. 2019;4(4): 1-8. DOI: 10.9734/AJRCS/2019/v4i430078

[11] Gyewali B, Maharjan B, Rana G, Pandey R, Pathak R, Poudel PR. Effect of different organic manure on growth, yield and quality of Radish (*Raphanus sativus* L.). SAARC Journal of Agriculture. 2020;18(2):101-114. DOI: https://doi.org/10.3329/sja.v18i2.51112

[12] Aiso I, Inoue H, Seiyama Y, Kuwano T. Compared with the intake of commercial vegetable juice, the intake of fresh fruit and komatsuna (*Brassica rapa* L. var. perviridis) juice mixture reduces serum cholesterol in middle-aged men: a randomized controlled pilot study. Lipids in Health and Disease. 2014,13, 102. DOI: 10.1186/1476-511X-13-102

[13] Japan CROPs [Internet] 2020. Komatsuna. Available from: https://japancrops.com/en/crops/komatsuna/. [Accessed:2021-06-28]

[14] Department of primary industries and fisheries, Queensland [Internet] 2007. Komatsuna. Available from: https://web.archive.org/web/20090823032916/http://www2.dpi.qld.gov.au/horticulture/5308.html. [Accessed: 2021-05-15]

[15] Collett MG, Stegelmeir BL, Tapper BA. Could nitrile derivatives of

turnip (*Brassica rapa*) glucosinolates Be Hepato-or Cholangiotoxic in Cattle?. Journal of Agricultural and Food Chemistry. 2014;62(30); 7370-7375.

[16] Hiroyuki H, Endar H, Asmak A. Improving coffee husk compost quality. Journal of Nutrition and Dietetic Practice. 2020:4(1);001-009.

[17] Stratful I, Scrimshaw MD, Lester JN. Removal of struvite to prevent problems associated with its accumulation in wastewater treatment works. Water Environ Research. 2004:76;437-443.

[18] Kabdaslı I, Tünay O. Nutrient recovery by struvite precipitation, ion exchange and adsorption from source-separated human urine. A review. Environmental Technology Reviewes. 2018:7; 106-138.

[19] Bhuiyan MIH, Mivinic DS, Beckie RD. A solubility and thermodynamic study of struvite. Environmental Technology. 2007: 28; 1015-1026.

[20] Le Corre KS, Valsami-Jones E, Hobbs P, Parsons SA. Impact of reactor operation on success of struvite precipitation from synthetic liquors. Environmental Technology. 2007: 28; 1245-1256.

[21] Harada H, Katayama Y, Afriliana A, Inoue M, Teranaka R, Mitoma Y. Effects of co-existing ions on the phosphorus potassium ratio of the precipitate formed in the potassium phosphate crystallization process. Journal of Environmental Protection. 2017:08(11):1424-1434.

[22] Alessio S, Carlo L, Giulia MC, Raffaele M. Advances in struvite precipitation technologies for nutrients removal and recovery from aqueous waste and wastewater. Review. Sustainability. 2020:12; 7538. https://doi.org/10.3390/su12187538

[23] Quintana M, Colmenarejo MF, Barrera J, García G, García E, Bustos A. Use of bioproduct of magnesium oxide production to precipitate phosphorus and nitrogen as struvite from wastewater treatment liquors. Journal of Agricultural and Food Chemistry. 2004:52; 294-299.

[24] Quintana M, Sánchez E, Colmenarejo MF, Barrera J, García G, Borja R. Kinetics of phosphorus removal and struvite formation by the utilization of by-product of magnesium oxide production. Chemical Engineering Journal. 2005:111; 45-52.

[25] Jaffer Y, Clark TA, Pearce P, Parson SA. Potential phosphorus recovery by struvite formation. Water Research. 2002:36; 1834-1842.

[26] Korchef A, Saidou H, Amor MB. Phosphate recovery through struvite precipitation by CO2 removal: Effect of magnesium, phosphate and ammonium concentrations. Journal of Hazardous Materials. 2011: 186; 602-613.

[27] Chauhan CK, Vyas PM, Joshi MJ. Growth and characterization of Struvite-K crystals. Crystal Research and Technology. 2011: 46(2): 187-194.

[28] Mathew BYM, Schroeder LW. 1979. Crystal structure of a struvite analogue, MgKPO4:6H20. Acta Crystallographica Section B. 1979:35: 11-13.

[29] Chi Z, Kangning Xu, Min Z, Jiyun Li, Chengwen W. Factors affecting the crystal size of struvite-k formed in synthetic urine using a stirred reactor. Industrial and Engineering Chemistry Research. 2018:57; 17301-17309.

[30] Caravaca F, Lax A, Albaladejo J. Organic matter, nutrient contents and cation exchange capacity in fine fractions from semiarid calcareous soils. Geoderma. 1999: 93; 161-196.

[31] Schjonning P, Christensen BT, Carstensen B. Physical and chemical

properties of a sandy loam receiving animal manure, mineral fertilizer or no fertilizer for 90 years. European Journal of Soil Science. 1994:45; 257-268.

[32] Zheng W, Zhang M, Liu Z, Zhou H, Lu H, Zhang W, Yang Y, Li C, Chen B. Combining controlled-release urea and normal urea to improve the nitrogen use efficiency and yield under wheat-maize double cropping system. Field Crops Research. 2016: 197; 52-62.

[33] Timilsena YP, Adhikari R, Casey P, Muster T, Gill H, Adhikari B. Enhanced efficiency fertilisers: A review of formulation and nutrient release patterns. Journal of the Science of Food and Agriculture. 2015: 95; 1131-1142.

[34] Jat RA, Wani SP, Sahrawat KL, Singh P, Dhaka S, Dhaka B. Recent approaches in nitrogen management for sustainable agricultural production and eco-safety. Archives of Agronomy and Soil Science. 2012: 58; 1033-1060.

[35] Geng J, Ma Q, Zhang M, Li C, Liu Z, Lyu X, Zheng W. Synchronized relationships between nitrogen release of controlled release nitrogen fertilizers and nitrogen requirements of cotton. Field Crops Research. 2015: 184; 9-16.

Section 2

Protein and Organic Compounds

Chapter 5

Recent Developments in the Crystallization of PLLA-Based Blends, Block Copolymers, and Nanocomposites

Amit Kumar Pandey and Shinichi Sakurai

Abstract

Despite the extensive studies of poly(L-lactic acid)(PLLA), the crystallization of PLLA-based materials is still not completely understood. This chapter presents recent developments of crystallization of PLLA-based blends, block copolymers and nanocomposites. The first section of the chapter discusses the acceleration of PLLA crystallization by the inclusion of biobased (solid and liquid state) additives. It was found that the solid state additives work as a nucleating agent while the liquid-state additive works as a plasticizer. Both type of the additives can significantly enhance the crystallization of PLLA, as indicated by crystallization half-time ($t_{0.5}$) values. Such composites are of great interest as they are 100% based on renewable resources. The second section talks about the enhanced formation of stereocomplex (SC) crystals in the PLLA/PDLA (50/50) blends by adding 1% SFN. It was found that the loading of SFN enhances the formation of SC crystals and it suppresses the formation of HC (homocrystal). The third section deals with confined crystallization of poly(ethylene glycol) (PEG) in a PLLA/PEG blend. The PLLA/PEG (50/50) blend specimen was heated up to 180.0°C and kept at this temperature for 5 min. Then, a two-step temperature-jump was conducted as 180.0°C → 127.0°C → 45.0° C. For this particular condition, it was found that PEG can crystallize only in the preformed spherulites of PLLA, as no crystallization of PEG was found in the matrix of the mixed PLLA/PEG amorphous phase. The last section describes the confined crystallization of PCL in the diblock and triblock copolymers of PLA-PCL. Furthermore, enantiomeric blends of PLLA-PCL and PDLA-PCL or PLLA-PCL-PLLA and PDLA-PCL-PDLA have been examined for the purpose of the improvement of the poor mechanical property of PLLA to which the SC formation of PLLA with PDLA components are relevant.

Keywords: Crystallization, poly(lactic acid), stereocomplex crystallization, poly (ethylene glycol), poly(caprolactone), biobased additives, improvement of crystallizability, X-ray scattering, crystalline block copolymer, crystalline polymer blend, confined crystallization

1. Introduction

Biobased polymers are gaining great popularity recently due to the increasing environmental concerns associated with conventional polymers. One such polymer is poly(lactic acid)(PLA), which is obtained from 100% natural resources such as corn starch and sugar cane. PLA has a good advantage of mechanical strength and modulus (comparable to PET), however, it has slow crystallization rate, low elongation at break, and processing difficulties due to the low thermal stability which significantly restricts its practical applications. PLA exists in three optical isomeric forms poly(L-lactic acid) (PLLA), poly(D-lactic acid)(PDLA), and poly(D, L-lactic acid) (PDLLA). The PLLA and PDLA both can be partially crystallized with a melting temperature of 170–180 °C. However, a racemic blend (50% L and 50% D) gives an amorphous polymer. Generally, commercial PLA grades are comprised of L-lactic acid in majority with small amount of D moiety. The thermal and mechanical properties of PLLA are significantly affected by the presence of D units in PLLA [1].

The study of crystallization behavior of PLLA is very important to control its thermal, mechanical, and gas-barrier properties. The crystalline structure of PLLA has been studied by many researchers [2–5]. It has been reported that the crystallization of PLLA leads to several crystal forms (α, α', β, and γ). The α form is the most stable polymorph which is developed from the melt or solution. The crystalline structure of PLLA α form is pseudo-orthorhombic with dimensions of a = 1.0683 nm, b = 0.617 nm, c (chain axis) = 2.78 nm, where the molecules adopt a 10_3 helical conformation. Aleman et al. [6] proposed the space group of $P2_12_12_1$ as the most plausible packing mode of 10_3 helices.

In this chapter, we review the recent developments [7–15] of crystallization of PLLA-based blends, block copolymers and nanocomposites. This chapter contains four sections. The first section deals with the enhancement in the crystallization of PLLA by adding biobased additives. Over the years, there have been several strategies employed by researchers to improve the crystallizability of PLLA [1, 16–19]. One of the most common and effective method is the addition of a nucleating agent. The nucleating agents are known to provide the sites for nucleation in polymers which results in the enhancement of overall crystallization process. Most of the nucleating agents reported for PLLA (talc, carbon nanotubes, graphene, clay) are inorganic materials that are non-biodegradable in nature [1, 20]. Recently, it is an emerging trend to utilize renewable resources for the improvement of crystallizability of PLLA. In this regard, we used solid-state biobased additives like silk fibroin nanodisc (SFN) and cellulose nanocrystal (CNC) with the aim of improving the crystallization of PLLA. The SFN is a biobased and environmentally benign material which was extracted from the waste of muga silk cocoons, which is composed of 83.8% poly(L-alanine) [21]. The CNC is also a biobased material which was extracted from the waste of marine green algae biomass residue (ABR). Further, we used liquid-state biobased additive, i.e., organic acid monoglyceride (OMG) for the sake of improvement of crystallizability of PLLA. The differential scanning calorimetry (DSC), polarizing optical microscopy (POM), synchrotron small-angle X-ray scattering (SAXS), and wide-angle X-ray scattering (WAXS) measurements were used for the study of crystallization of PLLA. It is worthy to mention here that the time-resolved SWAXS (simultaneous measurements of SAXS and WAXS) technique is one of the most promising technique to detect the initiation of nucleation and follow the change in the structure of growing crystals during the crystallization from the melt.

The second section talks about the stereocomplex crystallization of PLA. When PLLA (left-handed helix) and PDLA (right-handed helix) are mixed, the resultant mixture is known to form a complex so-called "stereocomplex (SC)". The SC is known to improve the thermal stability of PLA [22, 23]. This is due to the approximately 50°C higher melting temperature of the SC crystals compared to the PLLA or PDLA homopolymer crystal (HC). While pure PLLA and PDLA crystallize in pseudo-orthorhombic form with a 10_3 helix conformation, the SC has a 3_1 helix form [24]. The crystalline structure of PLA stereocomplex is triclinic with dimension of $a = b$ = 0.916 nm, c (chain axis) = 0.87 nm, $\alpha = \beta$ = 109.2°, and γ = 109.8°, in which PLLA and PDLA chains are packed parallel taking 3_1 helical conformation [25]. The formation of SC is influenced by the molecular weights of PLLA and PDLA. It is very challenging to get SC crystals exclusively in the high molecular weight PLLA/PDLA blend due to the competing formation of SC and HC during the crystallization [26]. In view of this, we added 1% SFN in PLLA/PDLA (50/50) blend, aiming to enhance the SC crystallization.

The third section deals with the blend of PLLA and poly(ethylene glycol), PEG. PEG is a biocompatible polymer which is known for improving the toughness of PLLA [20, 27, 28]. The crystallization study of the PLLA/PEG blend is important from the aspect of the structural development, due to the fact that both the component (PLLA and PEG) are crystallizable having different T_g and T_m. Since PLLA and PEG are known to be miscible with each other, PLLA/PEG blend has attracted many researchers for the studies of structure control. Although there have been extensive research on the crystallization of PLLA part in the dual crystalline PLLA/PEG blend [28–32], the effect of PLLA spherulites on the PEG crystallization is not well-known. In this regard, we studied the effects of space confinement when the PEG crystallizes from the molten mixed amorphous phase sandwiched by the crystalline lamellae of the PLLA.

The final section of this chapter deals with the block copolymers of PLA (PLLA or PDLA) and poly (ε-caprolactone), PCL. PCL is a biodegradable polymer which is also known for improving the toughness of PLLA [28, 33]. Since it is known that PLA and PCL are immiscible, copolymerization is a better route in comparison with the blending to avoid the macro-phase separation. In view of this, we studied the crystallization behavior of dual crystalline PLLA-*b*-PCL, PDLA-*b*-PCL diblock copolymers by changing the block length of PLLA or PDLA, however, the block length of PCL was fixed. Furthermore, the blend of PLLA-*b*-PCL and PDLA-*b*-PCL is also studied for the study of formation of SC crystal by changing the block length of PLLA and PDLA components.

The PLLA samples were obtained from NatureWorks and the PDLA was obtained from Purac. The sample characteristics are summarized in **Table 1**. The specimens preparation method is mentioned in the respective section.

The DSC measurements for the isothermal crystallization were performed by DSC214 *Polyma* (NETZSCH, Germany). The specimens were first melted at 200°C (or 260°C) for 5 min and immediately cooled to T_{iso} with the cooling rate of 300°C/min, and kept isothermally until the completion of the crystallization process. POM observations were conducted by using a Nikon Eclipse *Ci*-POL polarizing optical microscope equipped with the Linkam THMS600 hot stage (Linkam Scientific, UK). The specimens were sandwiched between two coverslips. Next, the specimens were melted on the hot stage, then quickly cooled (cooling rate = 150°C/min) to the isothermal crystallization temperature, and then kept isothermally until the completion of the crystallization process. The POM images were taken under crossed polarizers with a 530 nm optical retardation plate inserted in the optical path. The time-resolved SWAXS measurements were carried out at the beamline BL-6A of Photon Factory at the KEK (High-Energy Accelerator Research Organization) in Tsukuba, Japan. The wavelength of the incident X-ray beam was 0.150 nm. The T-jump experiments were

Sample code	Optical purity	Number-average molecular weight (M_n)	M_w/M_n
PLLA 4032D (D1.4)	98.6%	1.66×10^5	2.05
PLLA 2500HP (D0.5)	99.5%	1.74×10^5	2.22
PDLA D130	> 99.5%	1.41×10^5	2.03

Table 1.
Sample characterization.

conducted using a sample holder designed to allow for a quick T-jump (385°C/min). The details of the experimental set-up are reported elsewhere [7].

2. Improvement of PLLA crystallizability by biobased additives

2.1 Solid state additives (nucleation agents)

In this section, we report the effect of solid-state additives (SFN and CNC) on the isothermal crystallization of PLLA from the melt (200°C).

2.1.1 Silk fibroin nanodisc (SFN)

The SFN used in this study was extracted from wastes of the muga silk (*Antheraea assama*) cocoon [21]. The crystalline portion (the β sheets) of the silk fibroin was isolated by using the acid-hydrolysis method. The obtained extract comprises 83.8% L-alanine, and the well-defined disc-like nano particles were obtained (see **Figure 1** for the chemical structure of poly(L-alanine)). Such morphology and dimensions have been reported as the average diameter and thickness of ~45 nm and ~ 3 nm, respectively [21]. The detailed information about the preparation of SFN can be found in Ref. [21]. PLLA/ SFN specimens were prepared by the solution-casting method, using dichloromethane (DCM) as a solvent [8]. The specimens are labeled as D1.4/SFN(x) or D0.5/SFN(x), where the numbers after D denote the % of D moiety in PLLA, and x stands for % of SFN inclusion.

POM observations were conducted to observe spherulites and to evaluate the growth rate of spherulites and the nucleation density as a function of time. The specimens were melted on the hot stage at 200°C for 3 min, then quickly cooled (cooling rate = 150°C/min) to the isothermal crystallization temperature (T_{iso}) of 120°C, and then kept isothermally until the completion of the crystallization process. The representative images of the evolution of spherulites for the D1.4 neat and D1.4/SFN(1.0) specimens at 120°C as a function of time are shown in **Figure 2(a)** and **(b)**. First, negative spherulites were observed for both of the D1.4 neat and D1.4/SFN(1.0) specimens. As shown in **Figure 2(d)** the total number of spherulites increased ~4.7 times (from 9 to 42). In addition, the induction period was shortened from 101 to 39 s. However, the growth rate of the spherulites was unchanged (5 μm/min) by the addition of SFN. These results clearly show that SFN can enhance nucleation of PLLA. SFN is considered to provide sites for easy formation of PLLA nuclei.

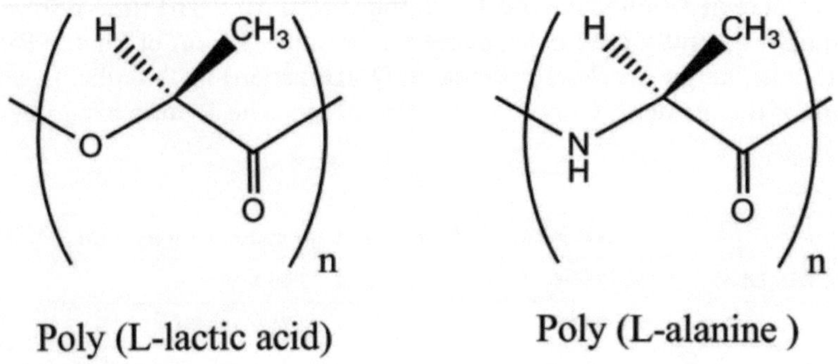

Figure 1.
Chemical structures of poly(L-lactic acid) and poly(L-alanine). L-alanine is the main component (83.8%) of the silk fibroin nanodisc.

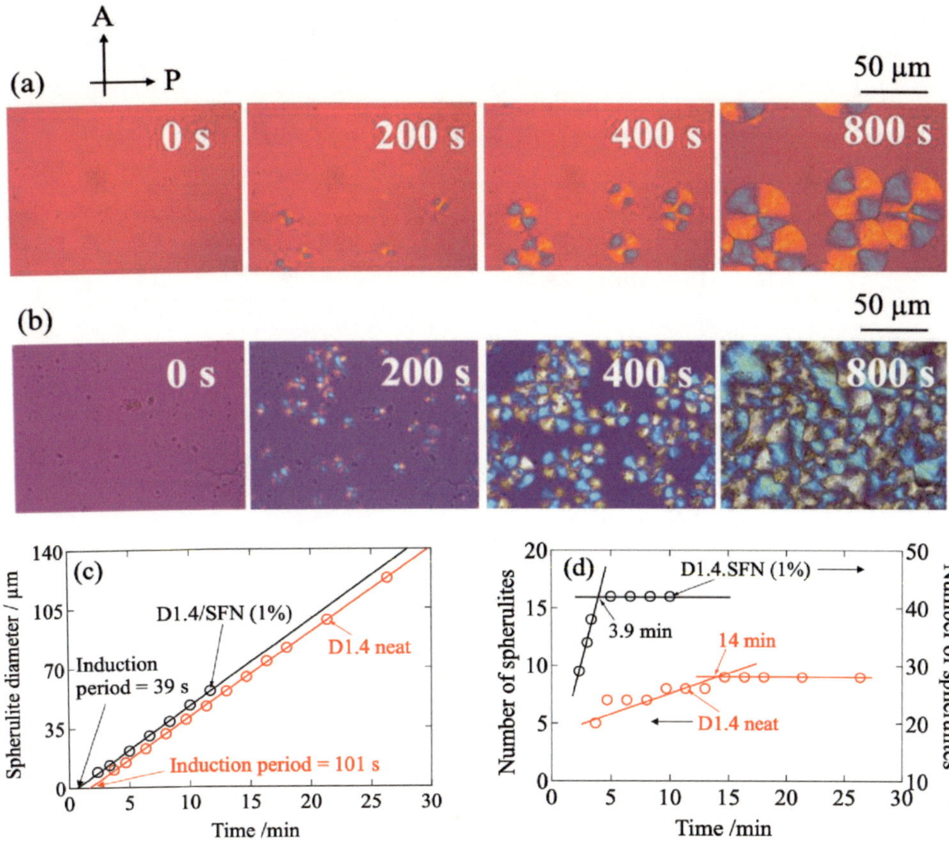

Figure 2.
POM images as a function of time for isothermal crystallization at 120°C (a) D1.4 neat (b) D1.4/SFN(1.0) specimens. (c) Spherulite diameter and (d) the number of spherulites formed during the isothermal crystallization at 120°C (adapted from reference [8] with a permission).

Figure 3 shows the DSC results of the isothermal crystallization of neat and 1% SFN included specimens at 110°C. the degree of PLLA crystallinity (ϕ_{DSC}) as evaluated based on the heat flow results using the following equation.

$$\phi_{\mathrm{DSC}}(t) = \frac{\int_0^t H(t)dt}{\Delta H_m^o} \tag{1}$$

where t denotes time and ΔH_m^o is the enthalpy of fusion for the 100% crystal of PLLA. The value of ΔH_m^o is taken as 93 J/g, following reference [34]. **Figure 3(b)** clearly indicates that the induction period was reduced and the final degree of crystallinity was increased by the presence of SFN. Maximum achievable crystallinity was found in the case of D0.5/SFN(1.0).

The inverse of crystallization half-time ($t_{0.5}$) can be used for the discussion of the crystallization rate, which is plotted as a function of the crystallization temperature in **Figure 4**. The graph shows a parabolic curve thereby producing maximum crystallization rate ($1/t_{0.5,\mathrm{max}}$). As shown in **Figure 4**, the overall crystallization growth rate was significantly increased by the inclusion of SFN, the maximum crystallization rate, $1/t_{0.5,\mathrm{max}}$, was observed at 107.2°C. It should be noted that for both specimens D1.4 and D0.5, the crystallization rate showed the same tendency, as the most effective temperature is ~107°C, although the T_m^O differs (The T_m^O

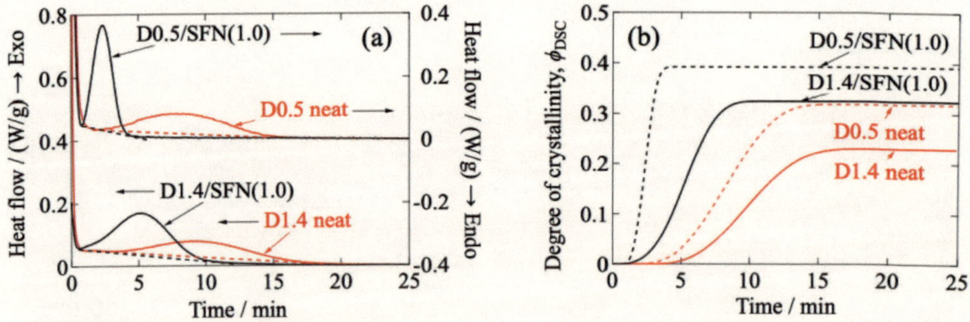

Figure 3.
(a) Heat flow as a function of time during isothermal crystallization at 110°C and (b) degree of crystallinity (ϕ_{DSC}) as a function of time, which was evaluated based on the heat flow result (adapted from reference [8] with a permission).

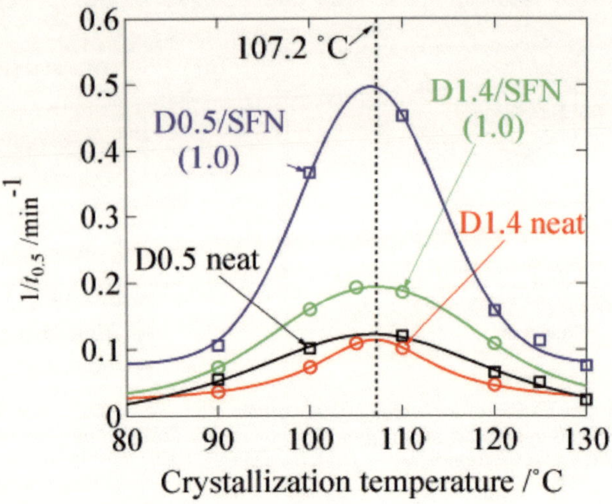

Figure 4.
Inverse of crystallization half time as a function of crystallization temperature, calculated from DSC results (adapted from reference [8] with a permission).

values for the D1.4 neat and D0.5 neat specimens were 180.7°C and 193.3°C, respectively [8]). These results indicate that the PLLA crystallization is predominantly governed by the kinetic driving force ($T - T_g$) rather than the thermodynamic driving force ($T_m^O - T$). It is important to note here that the spherulite growth rate does not change in the presence of SFN (**Figure 2**), while $1/t_{0.5}$ is increased. The reason for the enhancement of $1/t_{0.5}$ can be attributed to the increased number of nuclei due to the addition of SFN (**Figure 2**). The enhanced nucleation of PLLA by SFN may be ascribed to the plausible formation of hydrogen bonding between the C=O group in PLLA and the N–H group of poly(L-alanine) (see the chemical structure in **Figure 1**).

Figure 5(a) and **(b)** show the time-resolved WAXS profiles for the D1.4 neat and D1.4/SFN(1.0) specimens as a function of time for isothermal crystallization at 110°C. Here, the magnitude of the scattering vector $\boldsymbol{q}$ is defined as, $|\boldsymbol{q}| = q = (4\pi/\lambda) \sin(\theta/2)$ with λ and θ being the wavelength of X-ray and the scattering angle, respectively. As shown in **Figure 5(a)** and **(b)**, there is no crystalline peak initially, which shows the presence of 100% amorphous phase in the early stage. As time

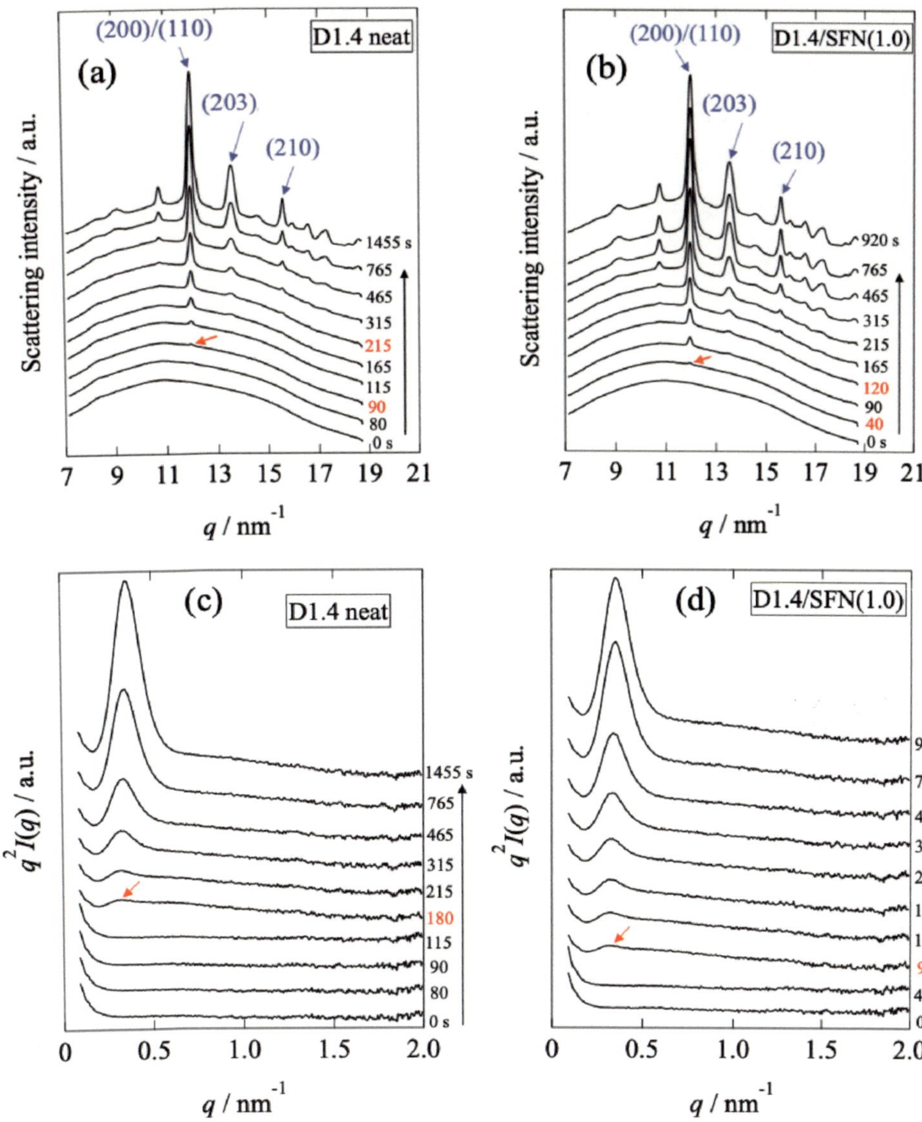

Figure 5.
Time-resolved (a)-(b) WAXS and (c)-(d) Lorentz-corrected SAXS profiles of D1.4 neat and D1.4/SFN(1.0) specimens. The red arrow indicates the first detection of the peak (adapted from reference [9] with a permission).

proceeds, a crystalline peak appears at q = 11.96 nm^{-1} (which has been shown by the red arrow). The induction period (t_0) of the crystallization is evaluated from the first detection of the crystalline peak. It was found that loading of 1% SFN decreased the t_0 from 90 s to 40 s, which shows that SFN enhanced the nucleation of PLLA.

The time evolution of the degree of crystallinity was calculated from the WAXS profiles by using the following equation

$$\phi_{\text{WAXS}} = \frac{\Sigma A_c}{\Sigma A_c + A_a} \tag{2}$$

Here, ΣA_c is the summation of the peak area of the crystalline peaks, and A_a is the peak area of the amorphous halo. The peak decomposition was conducted, and

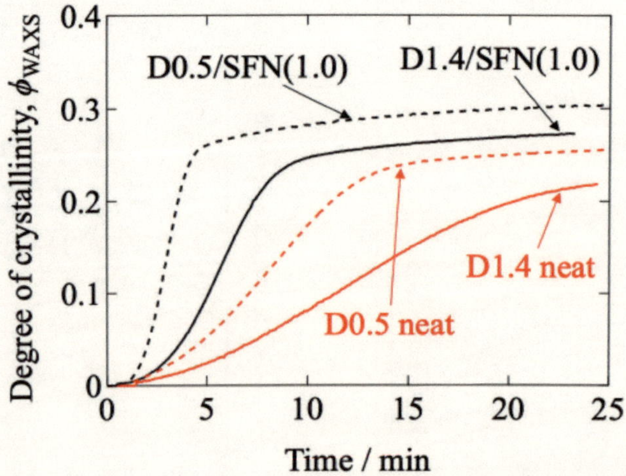

Figure 6.
Degree of crystallinity as a function of time for the isothermal crystallization at 110°C (adapted from reference [9] with a permission).

the degree of crystallinity ϕ_{WAXS} was calculated, which is plotted as a function of crystallization time in **Figure 6**.

As can be seen in **Figure 6**, the final degree of the crystallinity has been increased and $t_{0.5}$ is decreased by the inclusion of 1% SFN, indicating the acceleration of the crystallization rate.

Figure 5(c) and **(d)** show the changes in the Lorentz-corrected SAXS profiles as a function of time during the isothermal crystallization at 110°C for the D1.4 neat and D1.4/SFN(1.0) specimens. Here, the scattering intensity, $I(q)$, is corrected as $q^2I(q)$ by multiplying q^2. There was no SAXS peak observed in the early stage of crystallization. As time goes on, a clear scattering peak was observed at t = 180 s for the D1.4 neat or t = 90 s for the D1.4/SFN(1.0) specimen, which indicates the development of the lamellar stacking with sandwiching the amorphous layers. It is significant to observe that the SAXS peak appears later than the WAXS peak (**Figure 5**) which indicates that single lamellae (without stacking) are generated in the initial state of the PLLA crystallization from the melt.

As seen in **Figure 5(c)** and **(d)**, the SAXS peak moves towards the higher q as the crystallization proceeds. The long period (D) of the lamellar stacks was evaluated from the peak position (q^*) as $D = 2\pi/q^*$. As shown in **Figure 7(a)**, the D decreases as a function of the crystallization time which seemed to be conflicting to

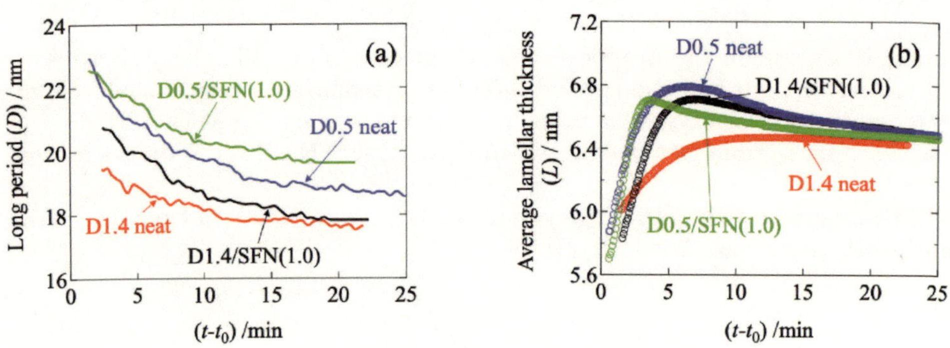

Figure 7.
SAXS results for isothermal crystallization at 110°C (a) long period (D), (b) average lamellar thickness (L) as a function of (t-t₀), where t₀ is the induction period (adapted from reference [9] with a permission).

the process of crystallization. To understand this behavior, the average thickness of the crystalline lamella (L) was calculated from the SAXS profile through the correlation function [35]. The correlation function is expressed as:

$$\gamma(r) = \frac{\int_0^\infty I(q)q^2 \cos(qr)dq}{\int_0^\infty I(q)q^2 dq} \tag{3}$$

Here, $\gamma(r)$ is the correlation function and r is the distance in the real space. $\gamma(r)$ function was obtained from the comprehensive $I(q)$ from $q = 0$ to $q = \infty$ by conducting the extrapolation of SAXS data for $q \to \infty$ according to Porod's law and for $q \to 0$, Guinier's law is used. The detailed procedure is reported in Ref. [7]. **Figure 7(b)** shows thus-evaluated L as a function of time. As a result, the average lamellar thickness, L increases with time, which is reasonable as a crystallization behavior. Therefore, the decreasing behavior of D (as shown in in **Figure 7(a)**) is also reasonable, as schematically shown in **Figure 8**. Upon crystallization, shrinkage takes place. Since the lamella thickens with time, this results in the decrease of D (**Figure 8(b)** and **(c)**), as the amorphous layer thickness is decreased to a greater extent as compared to the increasing extent of L (lamellar thickness).

2.1.2 Cellulose nanocrystal (CNC)

In this section, we discuss the enhancement in PLLA crystallizability by the inclusion of marine green algae biomass residue (ABR) based additives, i.e., washed ABR (WABR) and the ABR extracted cellulose nanocrystal (CNC). The CNC was extracted from the waste of ABR by using acid hydrolysis method. The complete extraction and characterization process is reported in Ref. [14]. Apart from effect of CNC on the crystallization behavior of PLLA, we also compare the utility of waste ABR after washing, i.e., WABR (washed algae biomass residue) as a filler for PLLA. As reported

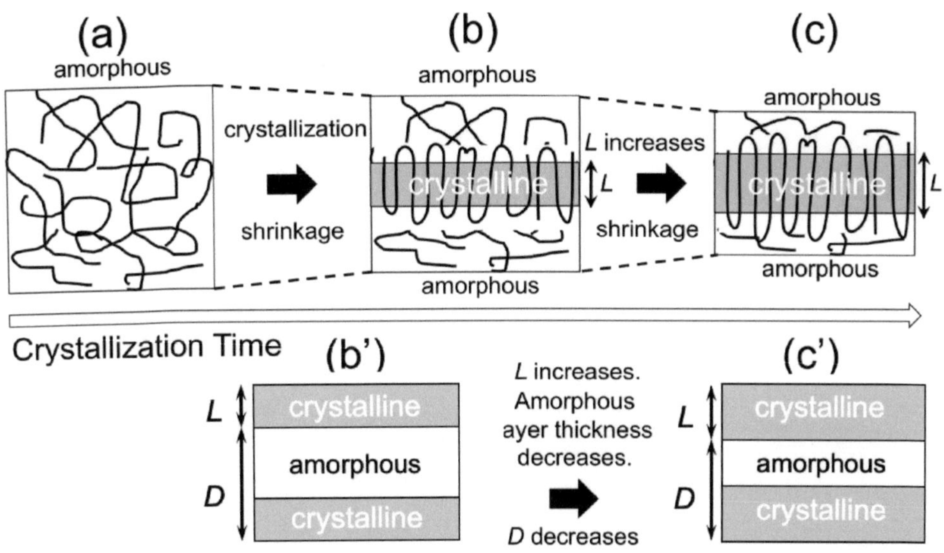

Figure 8.
Schematic illustrations showing the change in the nanostructure upon crystallization of PLLA. (a) At the amorphous state before crystallization of the polymer melt, (b) in an early stage of crystallization, (c) lamellar thickening in the subsequent stage of the crystallization (adapted from reference [9] with a permission).

in Ref. [14] it was found that WABR had irregular morphology (micron size), while the CNC had needle-like morphology with an average diameter of 30–35 nm, and average length of 520–700 nm. [14]. PLLA/WABR and PLLA/CNC composites were prepared by solution casting method using chloroform as a solvent. The loading amount of the additives were 0.5%, 1%, and 2% by weight. The effects of WABR and CNC on isothermal crystallization of PLLA are discussed by DSC and POM.

Figure 9 shows the degree of crystallinity as a function of time based on DSC results for the isothermal crystallization of neat PLA, PLA/WABR and PLA/CNC nanocomposites at 110°C. The degree of crystallinity (ϕ) was calculated by using Eq. (1). As shown in **Figure 9**, the addition of WABR, and CNC can improve the crystallizability of PLLA by reducing induction period, crystallization half-time, and by increasing the ultimate degree of crystallinity. It was observed that the CNCs were more effective as a crystallizing agent in comparison to WABR. Based on the DSC measurement in heating scan, it was found that WABR and CNC does not change the T_g and T_m of PLLA [14]. These results are similar to the case of loading SFN which is also a solid state additive [8].

Figure 10 shows the representative POM images for the isothermal crystallization at 125 °C for the neat PLA, PLA/WABR(1.0), and PLA/CNC(1.0) specimens at t = 12 min. First, negative spherulites were observed for all the specimens which shows that WABR and CNC had no effect on the structure of the PLA spherulites.

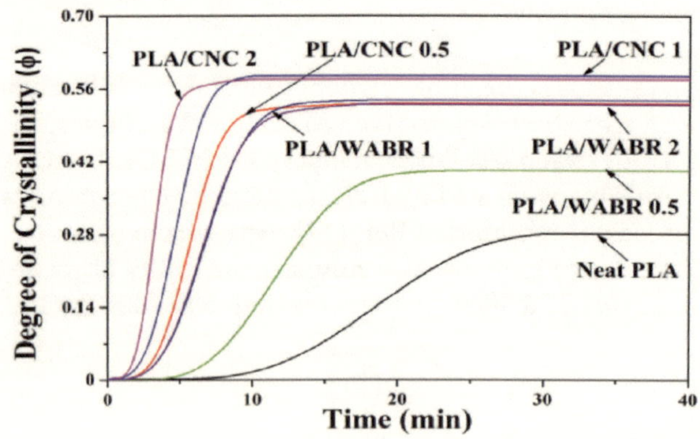

Figure 9.
Degree of crystallinity as a function of time based on DSC results for the isothermal crystallization of neat PLA, PLA/WABR and PLA/CNC nanocomposites at 110°C. the degree of crystallinity (ϕ) was calculated by using Eq.(1) (adapted from reference [14] with a permission).

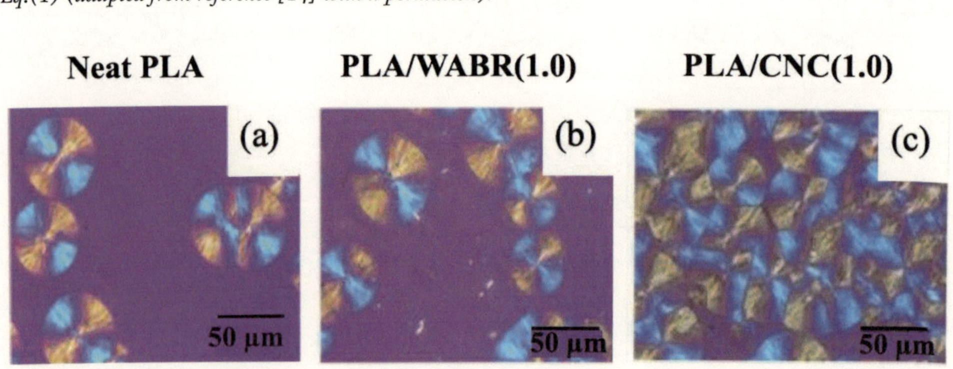

Figure 10.
The representative POM images for the (a) neat PLA, (b) PLA/WABR(1.0), and (c) PLA/CNC(1.0) specimens for the isothermal crystallization at 125°C for t = 12 min (adapted from reference [14] with a permission). The figure has been slightly modified.

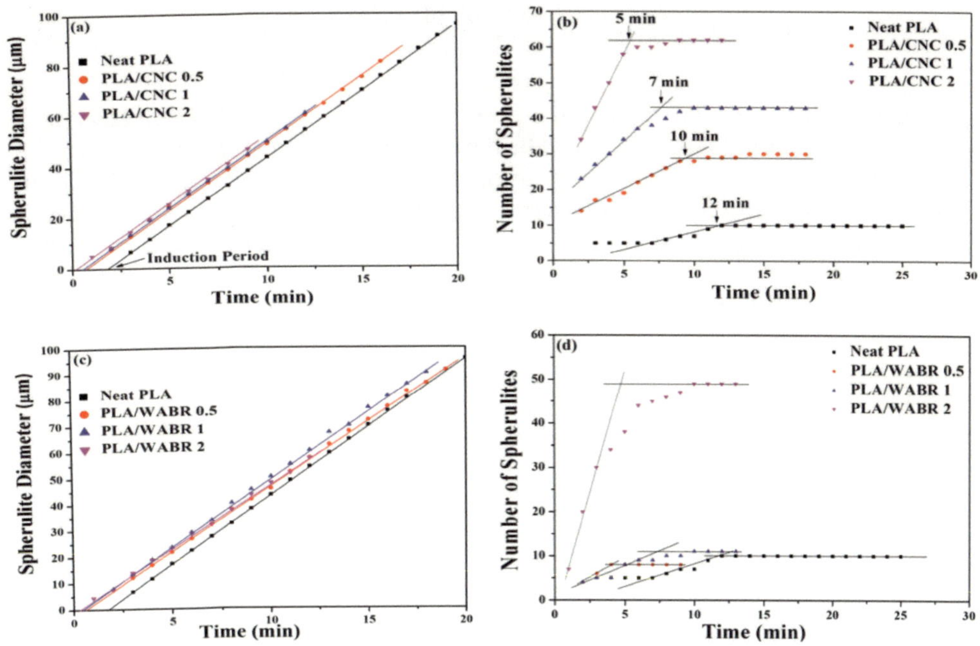

Figure 11.
(a) Spherulite diameter, (b) number of spherulites in the PLA/CNC bio-nanocomposites, and (c) spherulite diameter, (d) number of spherulites in the PLA/WABR bio-composites as a function of time at an isothermally crystallized temperature of 125°C (adapted from reference [14] with a permission).

Furthermore, **Figure 11** shows the spherulite diameter, and number of spherulites as a function of time, which were calculated from the POM images. It was observed that the incorporation of WABR, and CNC into the PLA matrix accelerated the rate of nucleation, however, the growth rate of the PLA spherulite was almost unchanged. CNC was found to be more effective than the WABR. The needle-like morphology and high aspect ratio of CNC were mainly accountable for the better effectiveness on improvement in the crystallization of PLA. On the contrary, the larger particle size of WABR might be the possible reason for its less effectiveness.

The results shown in this section suggest that even the low loading amount of solid state additives can enhance the crystallization of PLLA by providing more sites for nucleation without altering T_g, T_m, and spherulite growth during the isothermal crystallization from the melt.

2.2 Liquid state additive (plasticizer)

In this section, we will focus on the enhancement in crystallizability of PLLA by using a special plasticizer (organic acid monoglyceride; OMG). The chemical structure of OMG is shown in **Figure 12**. OMG is a product of Taiyo Kagaku Co., Ltd. The commercial name of OMG is Chirabazol D, which is a biobased plasticizer. The OMG has a molecular weight of 500 and a melting temperature of T_m = 67°C. PLLA/OMG specimens were prepared by the solution casting method, using chloroform as a solvent. The specimens are labeled as D1.4/OMG(x) or D0.5/OMG(x), where the numbers after D denote the % of D moiety in PLLA, and x stands for % of OMG inclusion.

Figure 13 shows the effect of OMG on the glass transition temperature (T_g) of PLLA. It is noticeably observed that T_g of PLLA decreases with the OMG content. More rigorously, we compare the result with the estimation by the plasticizing effect. The T_g can be simply estimated as.

$$T_g\ (\text{PLLA/plasticizer})(\text{K}) = \left(1 - \frac{\text{wt\%of plasticizer}}{100}\right) \times T_g(\text{neat})(\text{K}) \qquad (4)$$

Figure 13 shows the experimental T_g and the estimated T_g (with two straight lines) as a function of OMG content. The data points are almost in accord with the lines up to 1.0%, and then deviated from the lines above the OMG content of 1.0%. These results suggest that the OMG acts as a plasticizer for PLLA when the OMG loading is <1.0%. **Figure 14(a)** shows the DSC results of the isothermal crystallization of neat and 1% OMG loaded specimens at 110°C. The reduction in $t_{0.5}$ in **Figure 14(b)** confirmed the enhancement in crystallization rate. **Figure 14(c)**

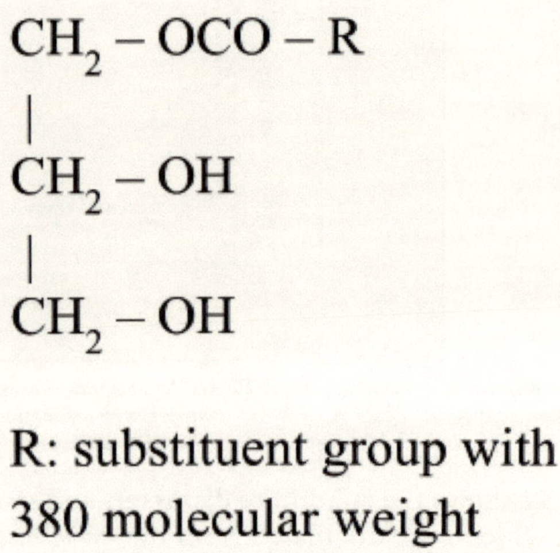

Figure 12.
Chemical structure of OMG.

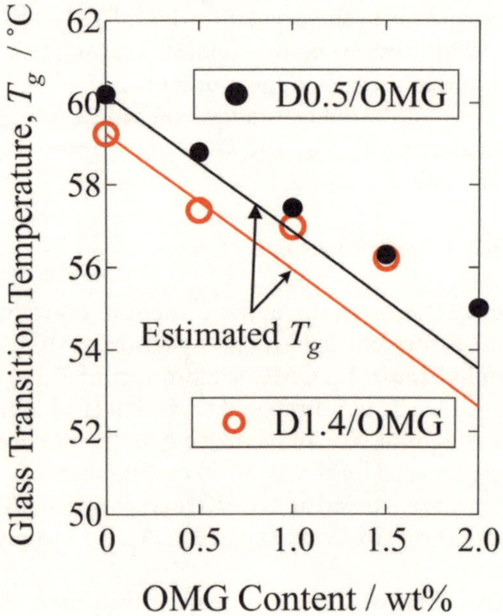

Figure 13.
Glass transition temperature (T_g) as evaluated from DSC curves. The lines show estimated T_g (adapted from reference [10] with a permission).

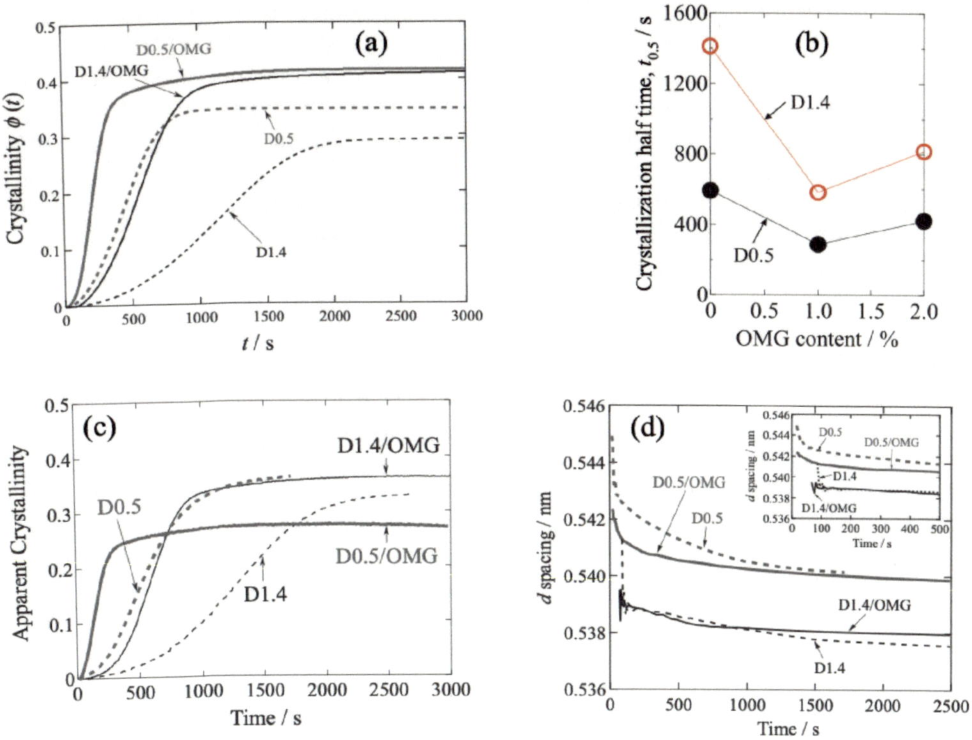

Figure 14.
(a) DSC results for isothermal crystallization at 110°C (b) crystallization half-time as a function of OMG loading (c) apparent crystallinity based on WAXS results, and (d) d spacing as a function of time (adapted from reference [7] with a permission).

shows the apparent degree of crystallinity which was evaluated using Eq. (2), based on the WAXS results as a function of the crystallization time for all specimens. The effect of the OMG is very clear for the acceleration of crystallization.

Figure 15(a) shows the change in long period, D as a function of time during the isothermal crystallization at 110°C for neat and 1% OMG loaded specimens. The long period decreases as the crystallization proceeds which can be explained by **Figure 8**. Furthermore, **Figure 15(b)** shows changes in the lamellar thickness (calculated from SAXS profile through the correlation function, $\gamma(r)$) as a function of

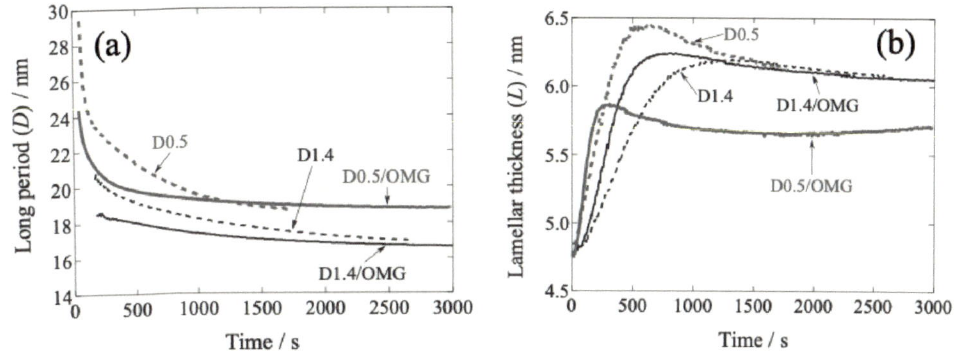

Figure 15.
SAXS results for isothermal crystallization at 110°C (a) long period (D), (b) lamellar thickness (L) as a function of time (adapted from reference [7] with a permission).

time. As time proceeded, the lamellar thickness increased in the early stage and then leveled off in the later stage. These results suggest that the lamellar thickness increased quickly in the early stage of crystallization due to a decrease in the D-content and the addition of OMG. The ultimate lamellar thicknesses (at 3000 s) for all specimens are relatively similar, although the value for the D0.5/OMG is 0.94 times those of the others.

The time-resolved Lorentz-corrected SAXS profiles during isothermal crystallization at 100°C form the melt (200°C) for D1.4/OMG and D0.5/OMG specimens are shown in **Figure 16**. There was no SAXS peak observed in the early stage. As the crystallization proceeds, the SAXS peak appears which gradually shifts towards the higher q range. As can be seen in **Figure 16(b)**, there was observed a clear second peak in the higher q range for the D0.5/OMG(1.0) specimen. However, the position of the new peak is not twice of the position of the first-order peak which means that the new peak is independent of the first-order peak. This result correspond to the newly formed lamellar stacking. There are three possible models to account for the appearance of a new peak in the higher q range. One is the formation of new lamellar stacks in the amorphous region with much shorter long period, as schematically shown in **Figure 17(a)**. The second one is the new lamellar stacks formed perpendicular to the original lamellar stacks, as schematically shown in the **Figure 17(b)**. This model is referred to as the cross-hatched lamellae [36–38]. The third one is the insertion of a new lamella into the amorphous phase, which is sandwiched by the neighboring two preceding lamellae, as schematically shown in the **Figure 17(c)**. This kind of insertion of a new lamella has been considered by

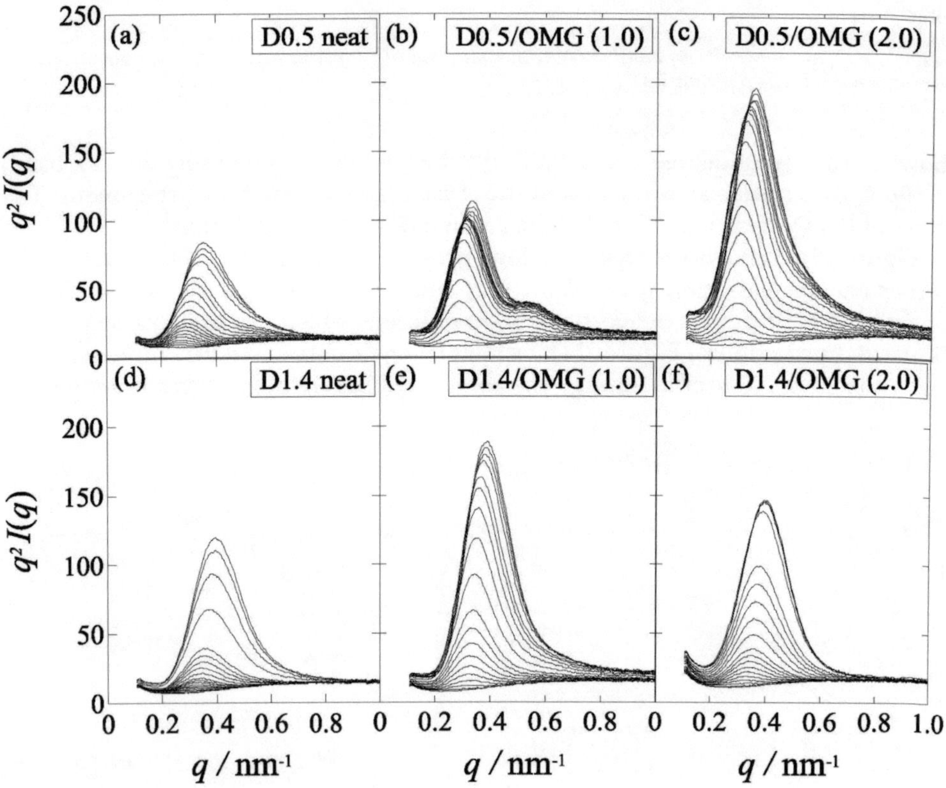

Figure 16.
Temporal changes in the Lorentz-corrected SAXS profiles upon T-jump from 200 °C to 100 °C for the specimen (a) D0.5 neat, (b) D0.5/OMG(1.0), (c) D0.5/OMG(2.0), (d) D1.4 neat, (e) D1.4/OMG(1.0), and (f) D1.4/OMG(2.0) specimens (adapted from reference [10] with a permission).

Figure 17.
Schematic representation for the formation of a new lamellar stack. (a) Independent stacking, (b) formation of a new stack in the amorphous phase in between the original lamellae with the stacking direction perpendicular to each other (cross-hatched lamellae) and (c) insertion of a new lamella in between the original lamellae (adapted from reference [10] with a permission).

Hama et al. [39]. At present, it is difficult to specify which model is appropriate. Although there was no such peak observed for the D0.5/OMG(2.0) specimen, the shoulders are clearly observed for this specimen (**Figure 16(c)**). Therefore, even for this specimen the effect of OMG to induce such a new lamellar stack can be recognized.

The POM observations were conducted to count the number of the spherulites as a function of time during the isothermal crystallization at 130 °C. **Figure 18(a)** and **(b)** show the representative POM images for the isothermal crystallization at 130 °C for the D1.4 neat and D1.4/OMG(1.0) specimens at t = 29 min. Since, the negative spherulites were observed which indicates that there is no effect of OMG on the structure of the spherulites. **Figure 18(b)** shows enhanced number of spherulites by loading of OMG which indicates that OMG can enhance the nucleation process of PLLA. From **Figure 18(c)**, it can be seen that the OMG enhances the spherulite growth of PLLA which clearly shows the effect of OMG to improve the crystallizability of PLLA. We speculate that the lowering of the activation energy for the PLLA crystallization may be the main effect of the OMG [10].

3. Enhancement in stereocomplex crystallization of PLLA/PDLA blend

In this section, the PLLA/PDLA (50/50) blends were prepared by solution casting method. Firstly, the PLLA and PDLA solutions were separately prepared with a concentration of 5% (w/v), using dichloromethane (DCM) as a solvent. The SFN was dispersed in DCM by using the ultrasonication method as discussed in the reference [8]. The PLLA, PDLA solutions, and the SFN dispersion, all together were mixed in one glass vessel and stirred for 12 h. The loading of SFN was 1% with the weight ratio of PLLA, PDLA, and SFN as 49.5/49.5/1.0. After the mixing, the solution was poured into a Petri dish for solvent evaporation at RT. After complete evaporation of the solvent, the as-cast films were obtained which were further dried in a vacuum oven at 50°C for 24 h. The specimens are labeled as LD neat and

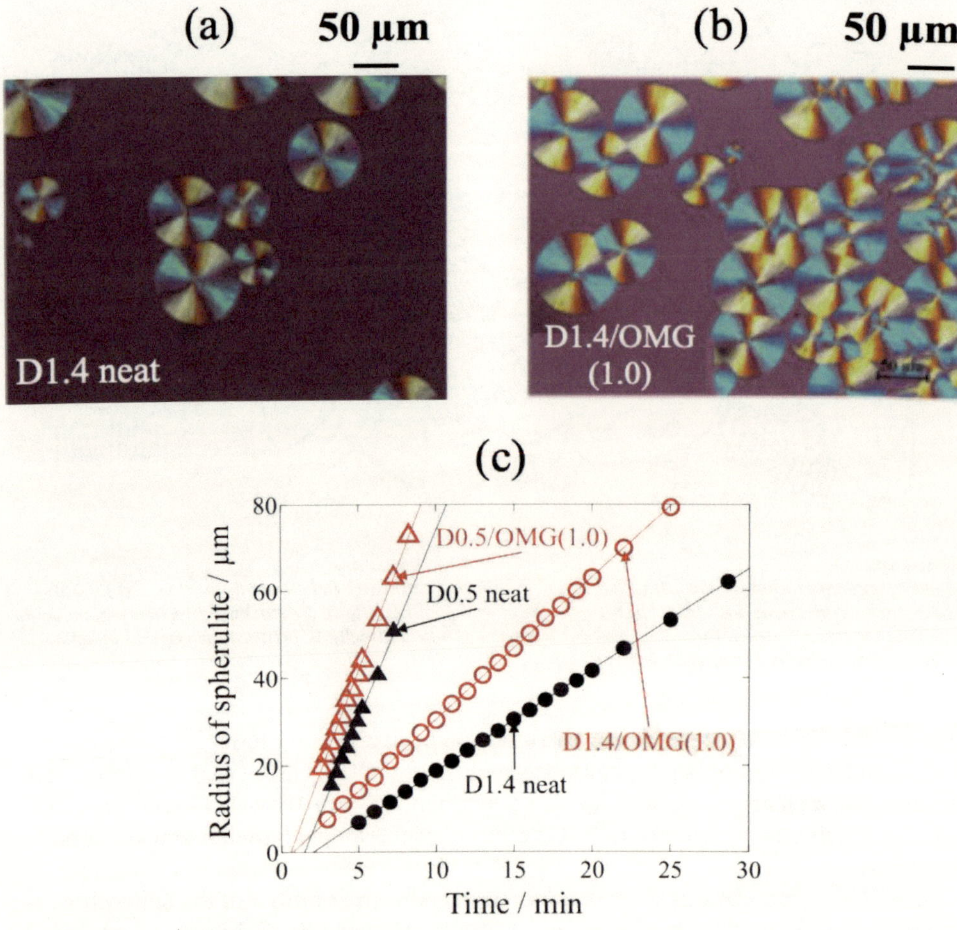

Figure 18.
(a), (b) POM images for the D1.4 neat and D1.4/OMG(1.0) specimens for the isothermal crystallization at 130°C for t = 29 min. (c) Plots of the radius of spherulite vs. time evaluated from the results of POM (adapted from reference [10] with a permission). The figure has been slightly modified.

LD/SFN(x), where LD denotes the blend of PLLA/PDLA(50/50), and x denotes the % loading of SFN.

Prior to the study of the effect of SFN on the crystallization of PLLA/PDLA (50/50) blend, we checked the effect of SFN on PDLA crystallization as SFN was known to improve the crystallization of PLLA (see Section 2.1.1). **Figure 19** shows the comparison of degree of crystallinity during isothermal crystallization of PLLA neat, PLLA/SFN(1.0), PDLA neat, and PDLA/SFN(1.0) specimens at 110°C. It can be seen from this figure that the ultimate degree of crystallinity (ϕ^{∞}) at the isothermal crystallization temperature of 110°C is increased by adding 1% SFN in PLLA or PDLA specimen. As shown in **Figure 19** the induction period, t_0 and the crystallization half-time, $t_{0.5}$ of the PDLA neat specimen are shorter than those of the PLLA neat specimen. This may be because the optical purity of the PDLA sample (D-content >99.5%) is higher than that of PLLA sample (L-content = 99.5%), as we know that the nucleation and crystallization of PLA (PLLA or PDLA) becomes quicker with the increasing optical purity. By adding SFN, the t_0 was almost unchanged for the case of the PDLA/SFN(1.0) specimen, while it was significantly decreased for the case of PLLA/SFN(1.0), furthermore, the $t_{0.5}$ is decreased for both cases. It was much decreased for the case of PLLA than that of PDLA, ensuring the

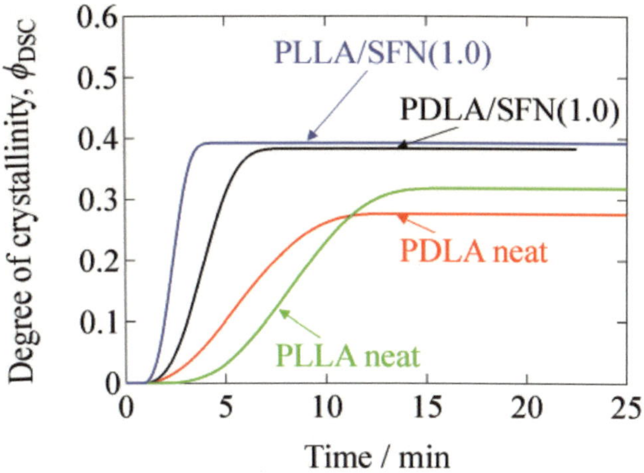

Figure 19.
Degree of crystallinity (ϕ_{DSC}) as a function of time, which was evaluated based on the heat flow results (adapted from reference [15] with a permission).

superior SFN effect due to its similarity of the chemical structure to PLLA. These results indicate the enhancement in the crystallizability of PDLA by adding 1% SFN, although SFN is much effective for the improvement of crystallizability of PLLA.

For the isothermal crystallization from melt, we set the melt temperature at 260° C for 5 min and then immediately quench to 110°C or 170°C and hold it isothermally until the crystallization completes. The reason why we selected 110°C is that it was found in **Figure 4** that the rate of crystallization of PLLA HC crystal is maximum at ~110°C. This is the best temperature to achieve the maximum amount of crystallinity of PLLA which is desirable for industrial purposes.

Furthermore, since at 110°C the formation of HC and SC occurs simultaneously so to see the effect of SFN on the formation of SC crystals solely, we conducted the isothermal crystallization at 170°C because at this temperature HC crystals cannot form ($T_{m,HC}$ = 170 ~ 180°C) due to the shallow quench depth ($\Delta T = T_m^O - T_c$).

The effect of SFN on the isothermal crystallization behavior of the PLLA/PDLA blend specimen was investigated at T_{iso} = 110°C. **Figure 20** shows the DSC results of the isothermal crystallization of LD neat and LD/SFN(1.0) specimens at 110°C from the melt (260°C). In **Figure 20(a)**, the heat flow as a function of time at the isothermal crystallization temperature is plotted. Adding 1% SFN, the crystallization exothermic peak shifts to the shorter time, showing an enhancement in the crystallization speed. However, it was not possible to distinguish the evolution of HC and SC phases from the plots of **Figure 20(a)**. To see the crystallites formed in the isothermal crystallization, the subsequent heating is conducted after the complete crystallization at 110°C. **Figure 20(b)** shows the results of the DSC heating scan. It is seen that by adding 1% SFN the $\Delta H_{m,HC}$ and $T_{m,HC}$ decreased and the $\Delta H_{m,SC}$ and $T_{m,SC}$ increased. These results indicate that the SFN can enhance the formation of SC and can suppress the formation of HC. The change in the melting temperature indicates that the presence of SFN may increase the lamellar thickness of the SC crystals while the lamellar thickness of HC crystals may be decreased due to the suppression of the HC crystallization. The increase in the melting point of SC is helpful to increase the thermal stability of PLA.

The effect of SFN on the isothermal crystallization behavior of the PLLA/PDLA blend specimen at 170°C was studied by the DSC measurement as shown in **Figure 20(c)**. Since the temperature 170°C is too high for the formation of HC ($T_{m,HC}$ = 170 ~ 180°C), only the formation of SC can be considered at this

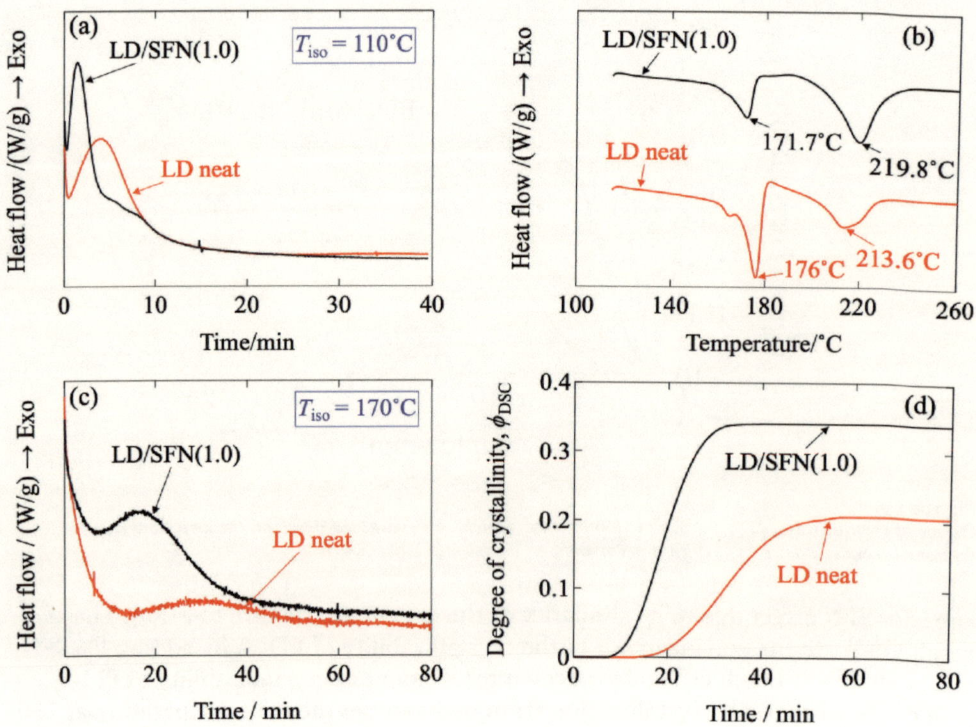

Figure 20.
(a) Heat flow as a function of time during the isothermal crystallization at 110°C from the melt (260°C), and (b) the subsequent heating scan from 110–260°C with the rate of 20°C/min. (c) Heat flow curves as a function of time during the isothermal crystallization at 170°C from the melt (260°C), and (d) changes in the degree of crystallinity (ϕ_{DSC}) as a function of time at 170°C, (adapted from reference [15] with a permission).

temperature (see later WAXS results in **Figure 21(c)** and **(d)**). From **Figure 20(c)** and **(d)**, it is clearly seen that the ultimate degree of crystallinity at the isothermal crystallization temperature of 170°C is increased by the presence of SFN. The t_0 was decreased from 13 min to 7.4 min and the $t_{0.5}$ was also decreased from 32.3 min to 19.1 min. These results indicate the enhancement in the stereocomplex crystallization of PLLA/PDLA (50/50) blend specimens by adding 1% SFN.

To clearly distinguish the evolution of HC and SC during the isothermal crystallization, we conducted the time-resolved WAXS measurements at 110°C upon T-jump from 260°C. **Figure 21(a)** and **(b)** show the change in WAXS profiles for the LD neat and LD/SFN(1.0) specimens as a function of time at 110°C. The peaks located at q = 8.75, 14.6, and 16.6 nm^{-1} belong to the SC crystals while the other reflection peaks belong to the HC. As shown in **Figure 21(a)**, even at the very early stage the $SC_{(110)}$ refection peak was observed for both the specimens, while the peak area of $SC_{(110)}$ was much larger for the case of LD/SFN(1.0) specimen. The shorter induction period of SC than HC is due to the fact that the nucleation of SC is faster than that of HC in PLLA/PDLA (50/50) blend by the difference in the thermodynamic driving force of the crystallization. ($\Delta T_{SC} > \Delta T_{HC}$ where $\Delta T_{SC} = T_{m,\,SC} -$ T and $\Delta T_{HC} = T_{m,\,HC} -$ T). As time goes on, the HC peak appears at 125 s. It is noteworthy to observe here that the induction period of HC is unchanged by adding SFN. The time evolution of the degree of crystallinity was calculated from the WAXS profiles which is plotted as a function of time in **Figure 22**. As can be seen from **Figure 22** for the case of LD neat specimen, the HC peak appears later than the SC peak while it keeps on increasing and finally, ϕ_{HC} overcomes ϕ_{SC}. For the case of LD/SFN(1.0) specimen, the SC crystallization is much accelerated

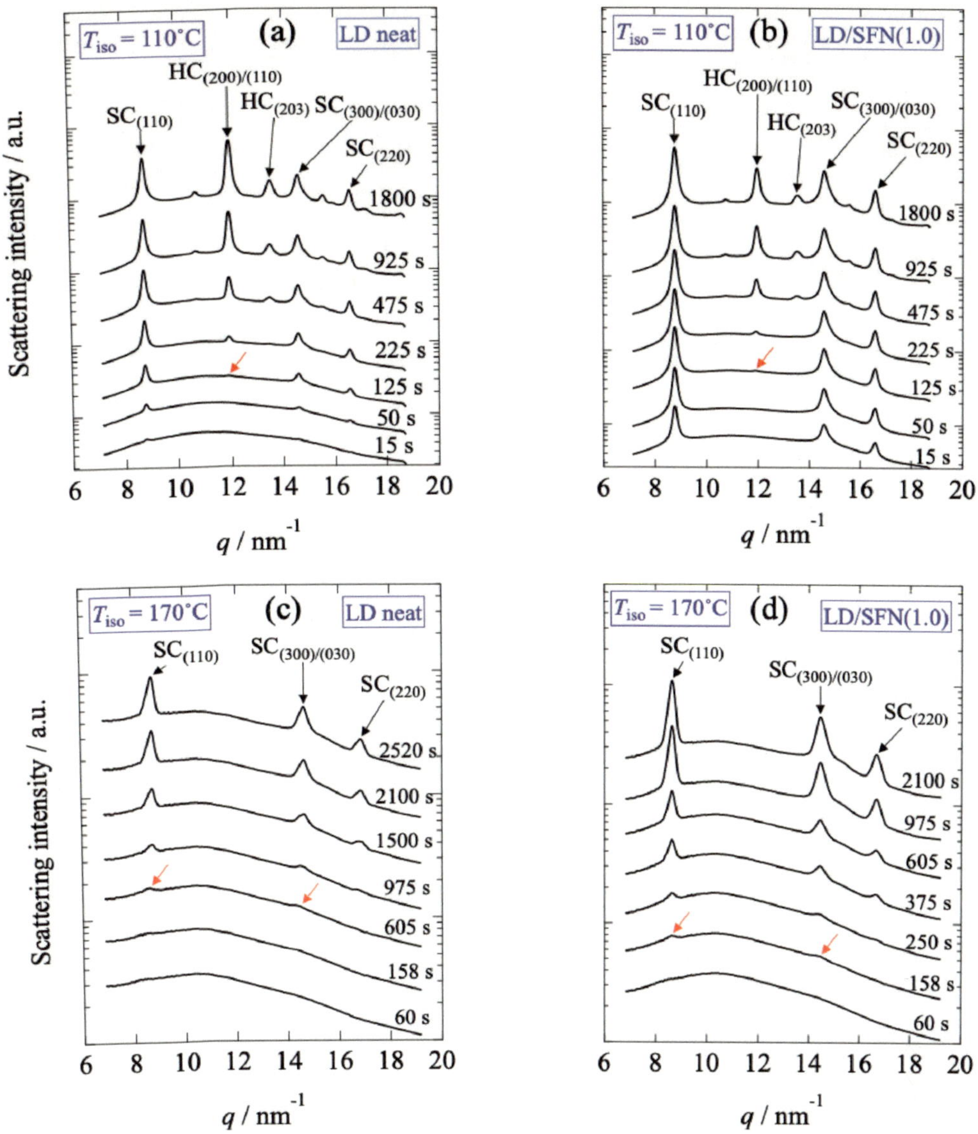

Figure 21.
Time-resolved WAXS profiles after the T-jump from 260–110°C for (a) LD neat (b) LD/SFN(1.0) specimens. (the red arrow indicates the first detection of the peak for the HC). (c) Time-resolved WAXS profiles after the T-jump from 260–170°C for (c) LD neat (d) LD/SFN(1.0) specimens. (the red arrow indicates the first detection of the SC peak) (adapted from reference [15] with a permission).

in the very early stage with the almost zero induction period and in the final stage $\phi_{HC} << \phi_{SC}$. The fraction of SC (f_{SC}) is increased after loading of 1% SFN while the total degree of crystallinity ($\phi_{HC} + \phi_{SC}$) is unchanged at 30 min.

The average crystallite size (D_{hkl}) in the direction normal to the (hkl) plane was evaluated by Scherrer's Equation [40].

$$D_{hkl} = \frac{K\lambda}{\beta_{hkl} \cos\left(\frac{\theta}{2}\right)} \tag{5}$$

where K is a constant (0.9) and λ is the wavelength of the incident X-ray. β_{hkl} is a full-width at half maximum (FWHM) in the unit of radian, and θ is the scattering

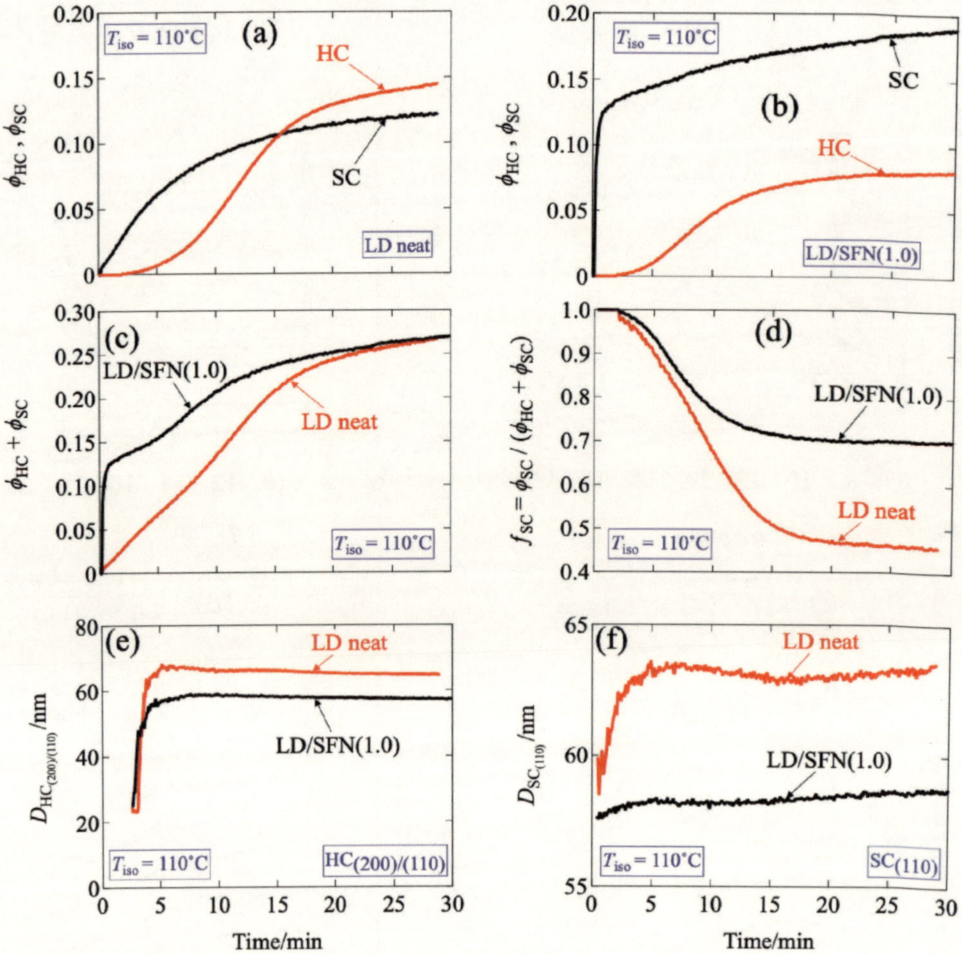

Figure 22.
Degree of crystallinity calculated from the results of ***Figure 21*** *for (a) LD neat and (b) LD/SFN(1.0) specimens. (c) Total (HC + SC) degree of crystallinity and (d) fraction of SC as a function of time. Plots of average crystallite size as a function of time evaluated by Scherrer's equation for (e) HC and (f) SC (adapted from reference [15] with a permission).*

angle. Note here that the raw data were used as the β_{hkl} values without correction for the peak broadening due to the collimation error of the WAXS setup, if any.

As seen in **Figure 22(e)** and **(f)**, the crystallite size is initially increasing as a function of time and it levels off after 5 min elapsed from the onset of crystallization. The slope of the plots in **Figure 22(e)** and **(f)** can be considered as the crystallite growth rate. Then, it can be stated that the growth rate of the HC crystallites is unchanged by the addition of SFN. The final value of the size of the HC crystallite for LD/SFN(1.0) specimens is slightly smaller than that in the LD neat specimen due to the effect of the SFN loading. As can be seen from **Figure 22(f)**, the size of the SC crystallite in the LD/SFN(1.0) specimen is much smaller than those of the LD neat specimen. Furthermore, it is interesting to notice that the initial size of the SC crystallite is the same for both the LD neat and the LD/SFN(1.0) specimens (**Figure 22(f)**).

To check the effect of SFN loading on the formation of SC crystals solely, we conducted the time-resolved WAXS measurements at 170°C. The changes in the WAXS profile were measured in the isothermal crystallization process at 170°C from the melt (260°C). **Figure 21(c)** and **(d)** show the WAXS profile for the LD neat and LD/SFN(1.0) specimens as a function of time. It is also clearly shown that there is no peak for HC crystals which is due to such a high temperature, i.e. 170°C. It can be said that at 170°C only SC crystal formation takes place.

Figure 23(a) and **(b)** show the changes in Lorentz corrected SAXS profiles as a function of time during isothermal crystallization at 110°C for the LD neat and LD/SFN(1.0) specimens. The SAXS profiles in **Figure 23(a)** and **(b)** show the scattering from both the HC and SC crystal (as evidenced by WAXS results in **Figure 21**). To distinguish the contribution of HC and SC crystal, we conducted the peak decomposition of the SAXS profiles. The detailed procedure about the peak

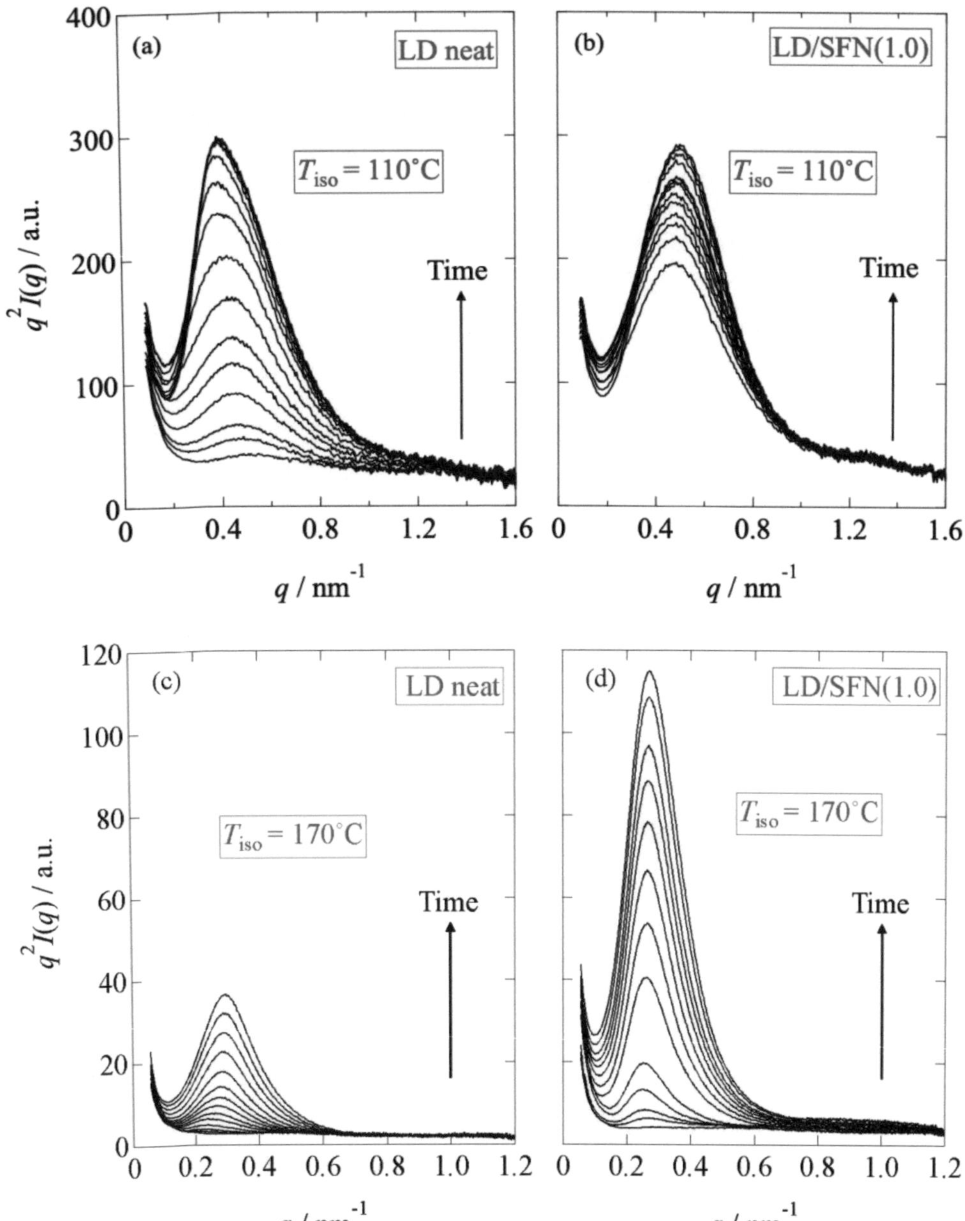

Figure 23.
Changes in the Lorentz-corrected SAXS profiles as a function of time for LD neat and LD/SFN(1.0) specimens during the isothermal crystallization at (a)-(b) 110° C and (c)-(d) 170°C from the melt (260°C) (adapted from reference [15] with a permission).

decomposition is mentioned in Ref. [15]. For $t < 6$ min, the SAXS profiles are symmetric which belong to SC crystals. As time goes on, after $t = 6$ min, the second peak appears which shows the scattering from HC crystals. From the peak position (q^*), the long period (D) of the lamellar stacks was evaluated as $D = 2\pi/q^*$. **Figure 24(a)** shows the plot of D as a function of time for the LD neat and LD/SFN (1.0) specimens which show the contribution of HC and SC separately. As seen in **Figure 24(a)**, D decreases as a function of time for HC crystals in the LD neat specimen. After loading 1% SFN, a similar trend was observed while the value of D of the HC lamellar stack was smaller than that of the LD neat specimen. The D of the SC lamellar stack in the LD neat specimen first increases up to $t = 6$ min and then decreases after $t > 6$ min. Considering **Figure 22(a)**, $t = 6$ min can be taken as $t_{0.5}$ (crystallization half-time). Then, the crystallization was quick in the stage $t < t_{0.5}$. Namely, D increased with time during the rapid crystallization while D decreased with time during the subsequent slow crystallization. **Figure 22(f)**, also suggests that the lateral size of SC lamellae was very rapidly increased for $t < 6$ min. Therefore, it can be considered that the SC lamellae grow in their lateral direction by folding the polymer chains in the amorphous region outside of the lamellar stacks. In the meantime, thickening of the lamellae can be considered during this stage ($t < t_{0.5}$). Namely, the lamellar thickening may be considered to take place by including the amorphous chains from outside of the lamellar stack. This situation quite differs from **Figure 8**, for which D is explained to be decreasing with time because of shrinkage in volume upon crystallization. As for the current case, no change in the amorphous layer with increasing of the thickness of the crystalline lamellae result in increasing the long period, D. For the LD/SFN(1.0) specimen D of the SC lamellar stack decreases from the beginning however the decreasing tendency became more evident for $t > 6$ min.

Figure 23(c) and **(d)** show the changes in Lorentz corrected SAXS profiles as a function of time during isothermal crystallization at 170°C for the LD neat and LD/SFN(1.0) specimens. The intensity of the peak observed at $q = 0.29\ \text{nm}^{-1}$ increases as a function of time. As seen in **Figure 23(c)** and **(d)**, the SAXS peak moves towards the higher q as the crystallization proceeds. Increase in q suggests the decrease in D as shown in **Figure 24(b)**. It can be seen that the long period, D decreases by the loading of SFN.

POM observations were conducted to evaluate the spherulite growth rate and the nucleation density as a function of time. The POM images of the evolution of spherulites for the LD neat and LD/SFN(1.0) specimens at 170°C as a function of

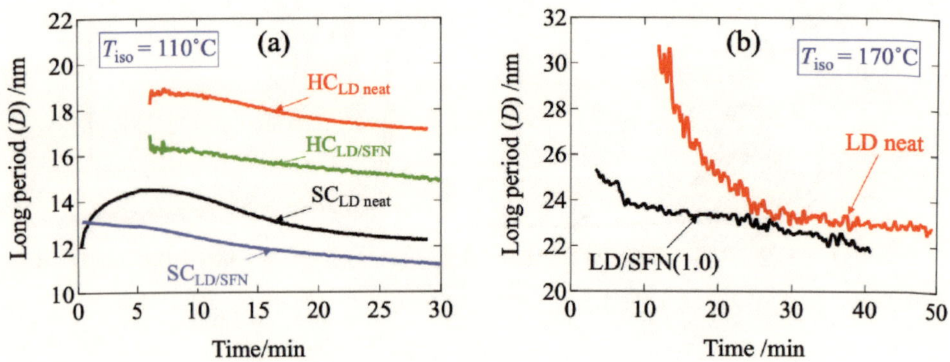

Figure 24.
Plots of long period (D) as a function of time during the isothermal crystallization at (a) 110°C and (b) 170°C from the melt (260°C) (adapted from reference [15] with a permission).

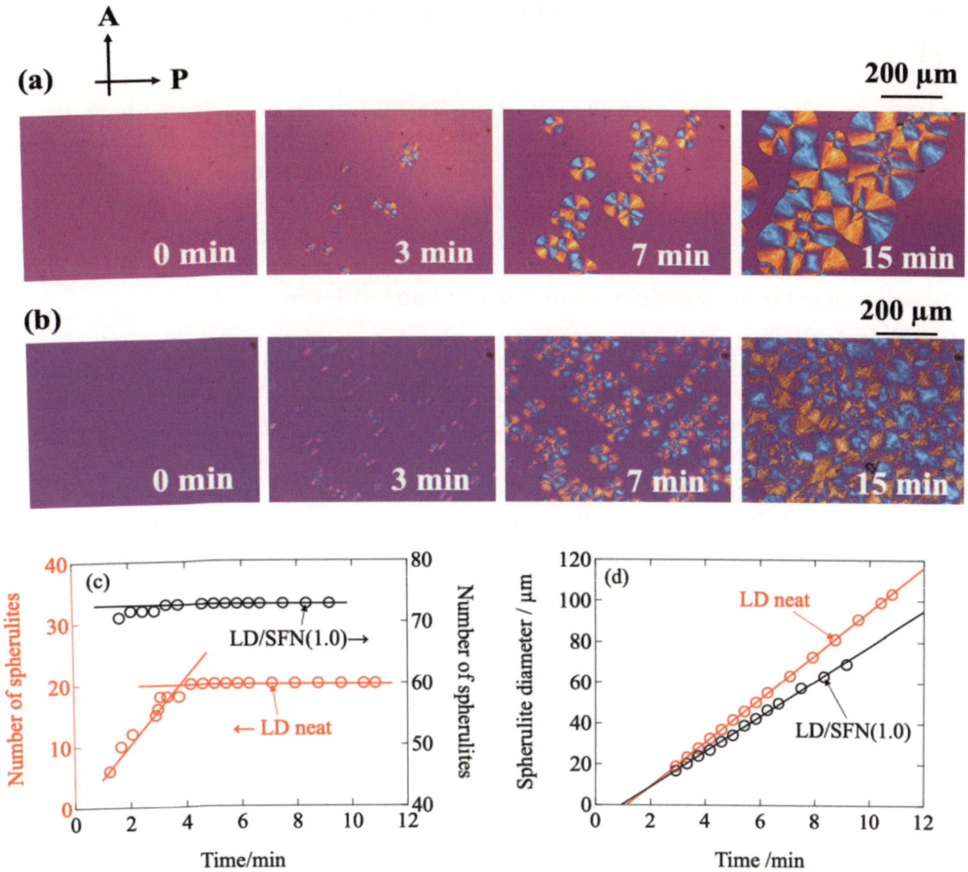

Figure 25.
POM images as a function of time for the isothermal crystallization at 170°C for (a) LD neat (b) LD/SFN (1.0) specimens. (c) Changes in the number of spherulites as a function of time, (d) the plots of spherulite diameter as a function of time during isothermal crystallization at 170°C, evaluated from the POM images (adapted from reference [15] with a permission).

time are shown in **Figure 25(a)** and **(b)**. First, negative spherulites were observed with the typical Maltese-cross patterns for both of the LD neat and LD/SFN(1.0) specimens. The number of spherulites and the spherulite diameter as a function of time are plotted in **Figure 25(c)** and **(d)**. As shown in **Figure 25(c)** the number of spherulites increases as a function of time for the LD neat specimen below 4 min, suggesting homogeneous nucleation. In contrast, for the case of LD/SFN(1.0), the number of spherulites significantly increases and kept constant as a function of time (**Figure 25(d)**), suggesting heterogeneous nucleation due to the nucleation effect of SFN. The final number of spherulites increased approximately 3.6 times (from 21 to 73) upon the addition of SFN. Based on these results, SFN is considered as a nucleation agent for SC nuclei. The induction period calculated from **Figure 25(c)** looks unchanged. Furthermore, as seen in **Figure 25(d)** the growth rate (8.6 μm/min) of the spherulites in the LD/SFN(1.0) specimen is smaller than that of the spherulites in the LD neat specimen (10.7 μm/min). Although the growth rate of the SC crystals is decreased by the loading of SFN, the ultimate degree of crystallinity at 170°C (see **Figure 20(d)**) is increased by the loading of SFN. The slower growth of SC spherulites by adding SFN seems conflicting with the larger nucleation effects of SFN. These two conflicting results (as shown in **Figure 25(c)** and **(d)**) induced by the SFN loading are worthy of future research.

4. Confined crystallization of PEG inside the preformed PLLA spherulite

In this section, we focus on the confined crystallization of PEG inside the preformed PLLA spherulite. The PLLA sample used in this study is the product of NatureWorks (code 4032D, D-content = 1.4%). The PEG sample was purchased from Wako Pure Chemical Industries, Ltd., of which M_w is 20,000. The PLLA/PEG (50/50) blend specimen was prepared by the solution casting method, using DCM as a solvent to obtain a solution with ca. 5 wt% of the total polymer concentration. The polymer solution was then poured into a Petri dish for complete evaporation of DCM.

The PLLA/PEG (50/50) blend specimen was heated up to 180.0°C and kept at this temperature for 5 min to obtain complete melt without liquid–liquid phase separation. Then, a two-step temperature-jump was conducted as 180.0°C → 127.0°C → 45.0°C. The isothermal crystallization time at 127.0 °C was controlled as 0, 5, 10, and 15 min where the PLLA spherulite grew. After that, the specimen was quenched to 45.0 °C and kept at this temperature for 40 min to induce the crystallization of PEG, as shown in **Figure 26**.

As can be seen from the POM micrographs in **Figure 27(a)**, PEG does not crystallize at 45.0 °C upon the direct quench from melt at 180.0 °C, however at 41°C PEG crystallization was clearly observed. This is due to the freezing temperature (T_f) depression of PEG in the mixture of the PLLA/PEG (50/50) blend specimen by noting that T_f for the neat PEG specimen is 52.0 °C. It should be noted that here we prefer to use the terminology "freezing temperature" instead of "crystallization temperature" to avoid any confusion with the "isothermal crystallization temperature" at which the isothermal crystallization experiment was conducted. Furthermore, the T_g of the PLLA/PEG (50/50) blend is approximately estimated as −8.2 °C by using the Fox equation with $T_{g,PEG}$ = − 53.0 °C [41] and $T_{g,PLLA}$ = 59.6 °C [8]. Therefore, the homogeneous mixture of PLLA and PEG is in the rubbery state at 41°C. However, PLLA crystallization was not observed at 41°C which may be due to the worse crystallizability of PLLA as compared to that of PEG.

The direct evidence of the confined crystallization of PEG inside the preformed PLLA spherulite was observed by the bright-field optical microscopic observation

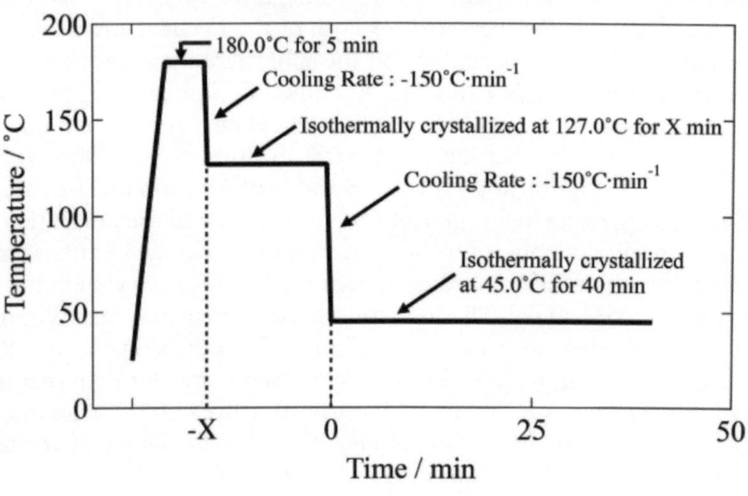

Figure 26.
Temperature protocol for the POM observations and the DSC measurements for the PLLA/PEG (50/50) blend specimens isothermally crystallized by the two-step T-jump (adapted from reference [13] with a permission).

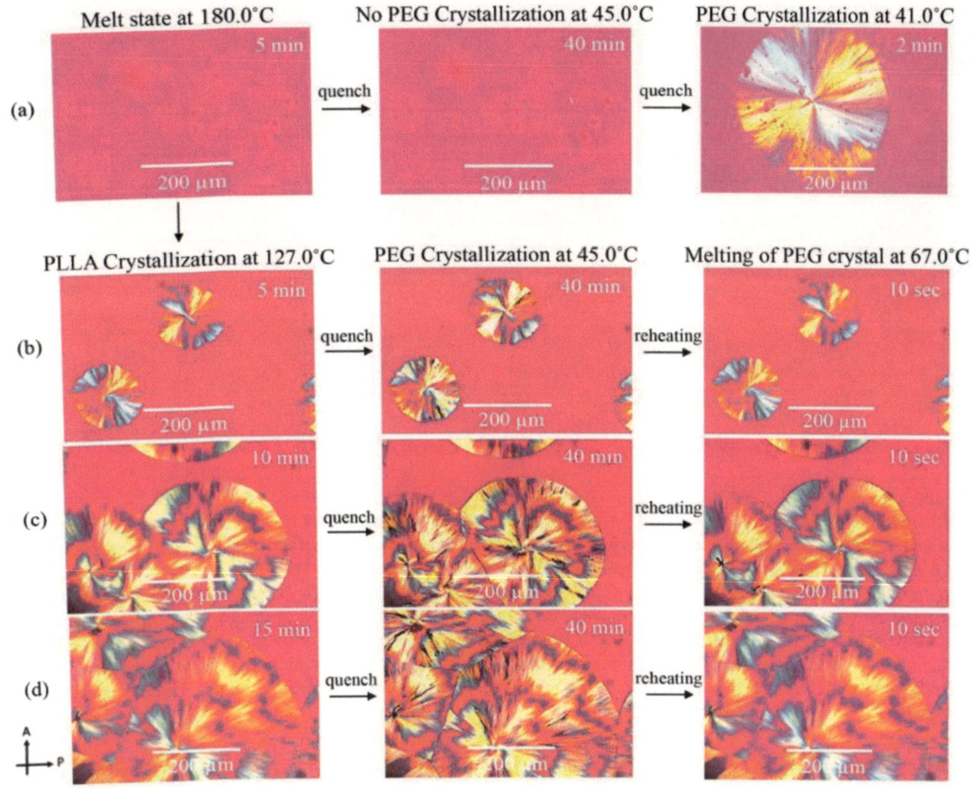

Figure 27.
POM micrographs showing the crystallization processes of PLLA/PEG (50/50) blend specimens at 45.0 °C (40 min) after the temperature jump from 127.0 °C, where PLLA was allowed to crystallize for (a) 0, (b) 5, (c) 10, and (d) 15 min. Subsequently, the specimens were subjected to reheating with 50°C min^{-1} up to 67.0°C. POM images shown in the right column were taken after 10 s of the temperature equilibration at 67.0°C (adapted from reference [13] with a permission).

which is shown in **Figure 28**. Actually, polarizer and analyzer plates were removed after first-step of T-jump at 127.0°C for 600 s. Afterwards, the specimen was quenched to 45.0°C. Around 484–486 s elapsed at 45.0°C, the dark spokes were observed inside the PLLA spherulite which were disappeared when temperature was increased up to 67.0°C. Thus, the confined crystallization of PEG in the preformed PLLA spherulite was evident. Upon further quenching from 67.0°C to 45.0°C, the confined crystallization of PEG again occured inside the PLLA spherulite, as shown in the bottom row of **Figure 28**. It is interesting to observe that the crystallization of PEG did not start from the center of the preformed PLLA spherulite. It rather seems that the initiation of the PEG crystallization was at random. Also, interesting to note no memory effect, i.e., the trajectories of the second-time PEG crystallization were completely different from the first-time ones. Furthermore, there observed a bridging PEG crystalline region which continuously strides over two-neighboring PLLA spherulites being contacted to each other with a straight boundary.

Figure 29(a) shows the change in the WAXS profiles during the isothermal crystallization at 45.0°C after the PLLA crystallization for 15 min at 127.0°C. Initially at t = 315 s, only the PLLA crystalline reflection peaks at q = 11.95, 13.43, and 15.45 nm^{-1} were observed. As time goes on, the PEG crystalline reflection peaks also appear at q = 13.56, 15.94, 16.12, 16.30 nm^{-1} with increasing their intensity. **Figure 29(b)**. shows the plots of peak area as a function of time. The peak area for

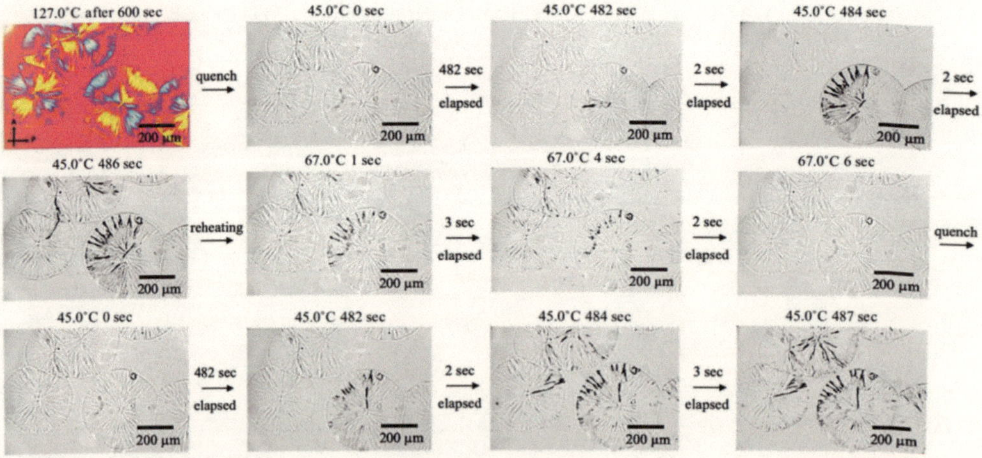

Figure 28.
POM image obtained at 127.0°C (600 s isothermal crystallization; shown at the top-left corner) and bright-field micrographs showing the confined crystallization of PEG in the PLLA spherulites for the PLLA/PEG (50/50) blend specimen. The specimen was quickly cooled from the melt state (180.0°C) to the crystallization state of PLLA at 127.0°C. after 600 s elapsed (isothermal crystallization for 600 s at 127.0°C), the specimen was again quenched to 45.0°C for the isothermal crystallization of PEG. Then, the specimen was subjected to reheating up to 67.0°C to melt the PEG crystalline phase. Subsequently (after 6 s elapsed), the specimen was quenched to 45.0°C for the isothermal crystallization at 45.0°C for the second time by keeping the same PLLA spherulites in which the confined crystallization of PEG took place again (adapted from reference [13] with a permission).

the PLLA crystalline peak was unchanged with the time however, that of PEG peaks showed the increasing tendency and then leveled off. The onset time of the peak evolution can be considered as the induction period which was about 5 min for this particular case. Although the peak positions of the PEG reflections in the second-step of T-jump of the blend specimen are the same as those for the neat PEG, it is specific to recognize the tremendous suppression of the $(12\bar{4})/(\bar{1}24)$ reflection peak for the blend specimen as compared to that for the neat PEG [13]. This may indicate the effect of the space confinement that the direction of the PEG crystal growth was suppressed, which is the $[12\bar{4}]/[\bar{1}24]$ direction almost parallel to the *c*-axis, in turn, the polymer chain direction. On the other hand, the (120) reflection peak was not suppressed, indicating that the PEG crystal growth in the direction perpendicular to the polymer chains was not affected. Assuming the folded-chain crystal of the PEG crystalline lamella, these results suggest the suppression of the lamellar thickening due to the space confinement in the amorphous phase sandwiched by the preformed PLLA crystalline lamellae. This further suggests the orientation of the PEG lamellae parallel to those of PLLA. The parallel orientation of the PEG lamellae (parallel to the preformed PLLA crystalline lamellae) as a consequence of the space confinement can be explained by the previous work of Huang et al. [42].

Such a space confinement effect results in the formation of extraordinarily thin PEG lamellae, in turn the lowering of the melting temperature according to the Gibbs–Thomson equation. To check this speculation, the DSC measurements were conducted. The specimens were first quenched from 180.0°C to 127.0°C to allow isothermal crystallization of PLLA for X min (X = 5, 10, 15, and 20) in prior to the second-step T-jump to 45.0°C to allow isothermal PEG crystallization at 45.0°C for 30 min. After the isothermal PEG crystallization at 45.0°C for 30 min, the specimen was then heated with the rate of 10°C/min where the DSC measurement was conducted. **Figure 30** shows the change in T_m of PEG as a function of the PLLA crystallization time at 127.0°C. It is clear that the T_m's of PEG in PLLA/PEG(50/50) blend are much lower than that for the neat PEG crystallized, suggesting the

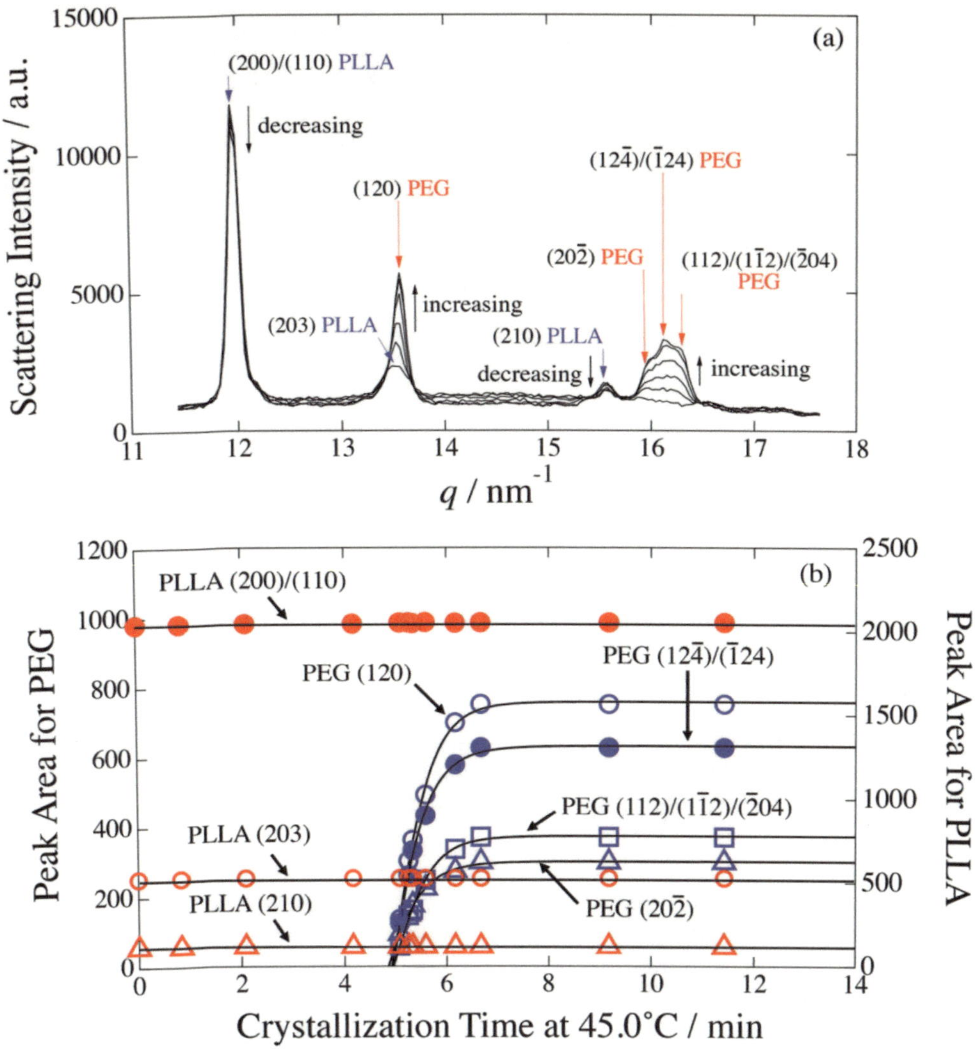

Figure 29.
(a) Time-resolved 1 d-WAXS profiles along the PEG crystallization at 45.0°C in the PLLA/PEG (50/50) blend specimen after PLLA crystallized at 127.0°C for 15 min. (b) Plots of the area of crystallization peaks as a function of time, which was evaluated from WAXS results shown in ***Figure 29(a)*** *after the peak decomposition (adapted from reference [13] with a permission).*

formation of thinner PEG lamellae in case of the confined crystallization. It is noteworthy to observe that T_m of PEG is monotonically increased with an increase of the PLLA crystallization time, which might imply that the space confinement effect becomes lesser with the growth of the PLLA spherulite and eventually reaching no space confinement effect for the PLLA crystallization time larger than 20 min. Although this tendency seems to be reasonable, it should be noted that thickening of the PLLA lamellae with an increase in the PLLA crystallization time results in more significant confinement to the PEG crystallization taking place in the amorphous region sandwiched by two PLLA crystalline lamellae. Therefore, the result shown in **Figure 30** rather implies the effect of the increase in the weight fraction of PEG (w_{PEG}) in the amorphous region comprising the homogeneous mixture of PEG and PLLA. Furthermore, there might be still a weaker confinement, as the density of crystalline PLLA (1.29 g/cm3) [43] is higher compared to that of amorphous PLLA (1.25 g/cm3) [43]. Consequently, at the same time as PLLA

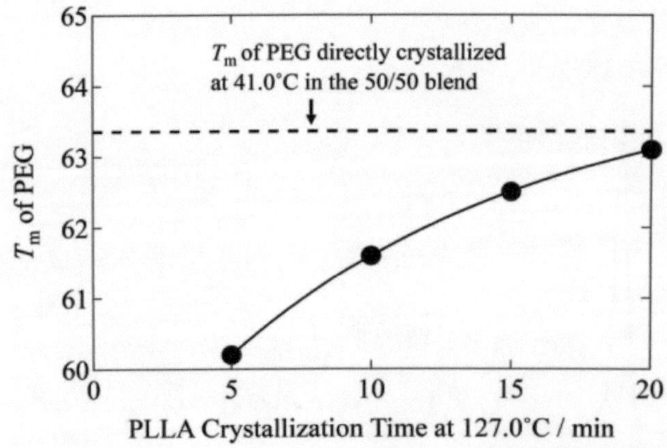

Figure 30.
T_m of PEG as a function of the PLLA crystallization time at 127.0°C (adapted from reference [13] with a permission).

lamellae grow in size, the amorphous phase will deplete in PLLA and due to the difference in densities there might be an overall gain in space, which enhances the growth of the PEG lamellae. As this effect becomes larger with the progress in the PLLA crystallization, the PEG lamellae can grow more so that the T_m of PEG increases with an increase in the PLLA crystallization time at 127°C, as shown in **Figure 30**. Thus, the DSC confirmation of the above-mentioned speculation of the formation of the PEG lamellae oriented parallel to the preformed PLLA lamellae is not satisfactory. We will report results of detailed DSC experiments using several PEG/PLLA blend specimens with different w_{PEG} in our future publication.

Furthermore, the T_f of PEG in the melt mixture of PEG and PLLA amorphous phase were determined by conducting the DSC measurements. Here, it is important to avoid crystallization of PLLA. Therefore, the specimens were quickly cooled from 180°C to 80.0°C and then cooled to 10°C with cooling rate 1,2 5 and 10°C/min. We conducted cooling-rate dependencies to correct for the cooling-rate effect on this temperature and to evaluate T_f of PEG by the extrapolation of the onset temperature of the exothermic peak to the zero-cooling rate [13]. Thus-evaluated T_f of PEG is plotted as a function of w_{PEG} for the PLLA/PEG blend specimens in **Figure 31(a)**. It can be seen that T_f is increased when w_{PEG} is increased. Based on this plot, the mechanism of the confined crystallization in the preformed PLLA spherulites is considered. Upon the crystallization of PLLA from the melt of the PLLA/PEG amorphous phase, the PEG content in the amorphous region inside the PLLA spherulite is increased so that T_f is increased from its original one (T_f = 45.3°C) at w_{PEG} = 0.5. The fact that no PEG crystallization at 45.0°C (**Figure 27(a)**) was observed for this blend specimen with w_{PEG} = 0.5 seems to conflict with the fact of T_f = 45.3°C. Since this value (T_f = 45.3°C) was estimated by the extrapolation of the onset temperature of the PEG crystallization to the zero-cooling rate, the PEG crystallization at 45.0°C would take infinitely long time for this blend specimen (PLLA/PEG (50/50)). **Figure 27(a)** showing no PEG crystallization at 45.0°C implies that the PEG crystallization would take place more than 40 min. Consequently, it can be considered that when the PLLA crystallization time at 127.0°C is longer, the PEG content in the amorphous region becomes higher so that the PEG crystallizability becomes more sufficient. The experimental results definitely supported this speculation.

Next, the PEG fraction in the amorphous phase in the preformed PLLA spherulite was estimated by DSC measurements. The PLLA/PEG (50/50) blend specimens were annealed at 180.0°C for 5 min, and then quenched first to 127.0°C for 10 min to form the PLLA spherulites. Afterward, the specimens were quenched to 60.0°C and then cooled gradually down to room temperature and DSC scans were observed [13]. Thus-evaluated T_f of PEG are plotted as a function of the crystallization time of PLLA at 127.0°C in **Figure 31(b)**. Based on the result shown in **Figure 31(b)** combined with **Figure 31(a)** it was possible to estimate w_{PEG}, which is increased from w_{PEG} = 0.5 with increasing of PLLA crystallization time at 127.0°C. **Figure 32** shows the Δw_{PEG} (= w_{PEG} – 0.5) behavior as a function of the PLLA crystallization time. It is clearly observed that w_{PEG} is increased with increasing of PLLA crystallization time at 127.0°C, which supports the above-mentioned discussion that w_{PEG} in the amorphous region inside the larger PLLA spherulite is larger than that inside the

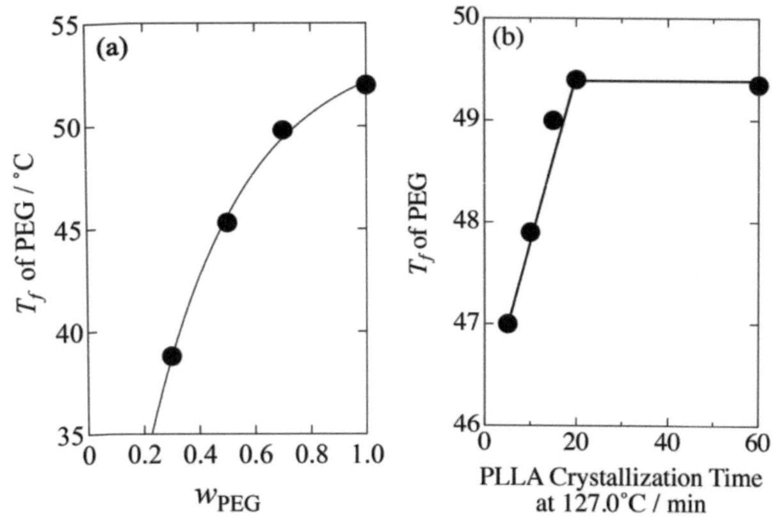

Figure 31.
(a) Dependence of the freezing temperature of PEG (T_f) on the weight fraction of PEG (w_{PEG}) in the PLLA/PEG blend specimens. (b) Dependence of T_f of PEG on the PLLA crystallization time at 127.0°C (adapted from reference [13] with a permission).

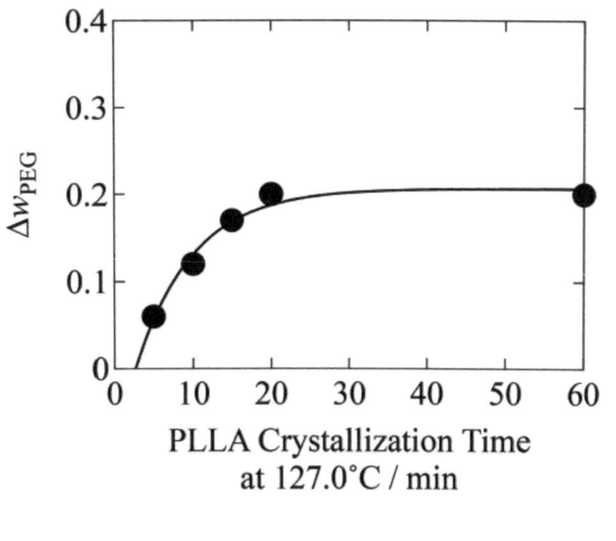

Figure 32.
Plot of w_{PEG} as a function of the PLLA crystallization time at 127.0°C (adapted from reference [13] with a permission).

smaller PLLA spherulite because of the progress in the PLLA crystallization. It is noteworthy that it leveled off around 20 min, suggesting completeness of the PLLA crystallization at 127.0°C around 20 min.

5. Confined crystallization of PCL in the block copolymer of PLA and PCL

In this section, we focus on block copolymers of PLA with PCL, namely, PLLA-PCL diblock copolymers and PLLA-PCL-PDLA triblock copolymers with respect to the confined crystallization within the microphase-separated domain of PCL which is sandwiched by the glassy PLLA microphase. In this study, PLLA, PDLA, PCL and their block copolymers were synthesized. The polymer synthesis method is described in Ref. [12]. The molecular weight information is reported in **Table 2**. The diblock copolymers are represented by XCL-YL and XCL-YD where X and Y denote the block length or the number-average molecular weights (M_n) in kDa of the component PCL (CL), PLLA (L), and PDLA (D) blocks. Whereas, the tri-stereoblock (trisb) copolymers are represented by XCL-YL-ZD where X, Y, and Z denote the block lengths or M_n values in kDa of its following block sequences shown by the abbreviated symbols. Further, the synthesized diblock copolymers were blended with their corresponding enantiomers, which are abbreviated as B_X-Y where X and Y denote the block length. For example, B_10–10 shows the blend of 10CL-10 L and 10CL-10D. Furthermore, all the prepared specimens were hot-pressed followed by quenching in ice-water, to obtain polymer films.

The DSC thermograms of the block copolymers compared with that of neat PCL and neat PDLA are shown in **Figure 33**. The enthalpy of melting (ΔH_m) of PCL decreases from 42.4 to 20.2 J/g with increasing the block length of PDLA, whereas the ΔH_m of PDLA is increased from 36.8 to 43.4 J/g. These changes in enthalpy correspond to the decreasing content of PCL and increasing content of PDLA in the copolymer system. Furthermore, different melting peaks of PCL and PDLA confirms separate crystallization of PCL and PDLA blocks of varying length. **Figure 33(b)** shows the DSC results of the enantiomeric diblock copolymer blends and the trisb copolymer in the heating scan (10°C/min). As can be seen in the figure, the specimen B_10–10 form perfect SC crystal upon blending, without generating the HC. This is because of the low molecular weight of the PDLA/PLLA blocks. Furthermore, double melting peaks were observed for SC crystal at 221.8°C and 235.4°C which may be due to the formation of relatively thinner and thicker lamellae of SC crystal. It should be noted that more formation of HC in B_10–20 and B_10_30 specimen is due to the increasing molecular weight of PLLA/PDLA blocks. Therefore, it can be concluded that the higher molecular weight of the block sequences increases the formation of HC instead of SC.

Specimen	M_n	M_w	M_w / M_n
10CL-10D	48.6	71.8	1.47
10CL-10 L	45.5	69.8	1.53
10CL-10 L-10D	59.0	100.6	1.70
10CL-20D	64.0	107.2	1.67
10CL-30D	69.8	160.5	2.30

Table 2.
Molecular weight of the synthesized block copolymers (analyzed by GPC).

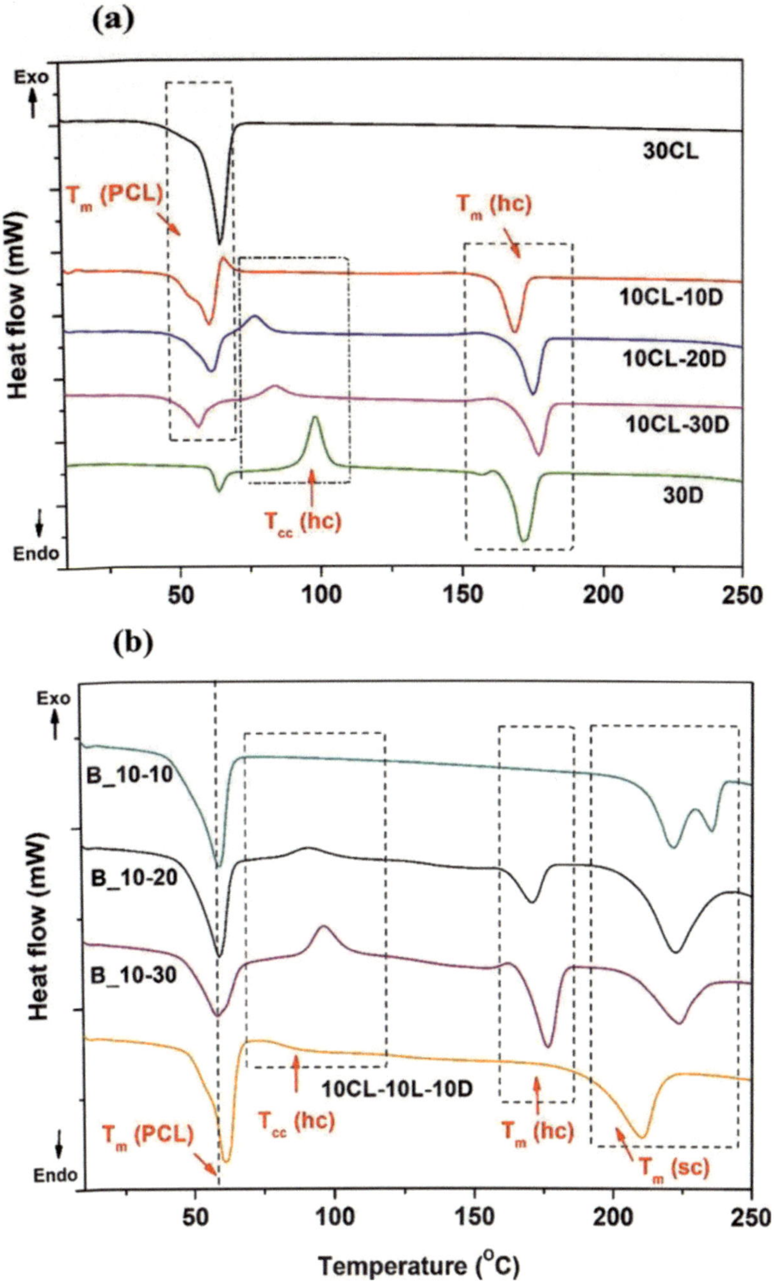

Figure 33.
DSC thermograms of (a) neat PCL, neat PDLA, and PCL-PDLA diblock copolymers and (b) enantiomeric diblock copolymer blends and trisb copolymer (adapted from reference [12] with a permission).

The stress–strain (SS) curves of the polymer films of the block copolymers and their enantiomeric blends as well as that of the trisb copolymer are shown in **Figure 34**. As can be seen that the elongation of 30PCL specimen is higher than that of 30D specimen. The higher elongation of PCL is due to its soft and flexible nature as the stress–strain test is performed at 25°C (rubbery region of PCL) while PDLA is in glassy state. The tensile strength, modulus, and the toughness of the diblock copolymers are enhanced with increasing the block length of PDLA. The highest elongation at break was found for the specimen 10CL-30D. To understand such an

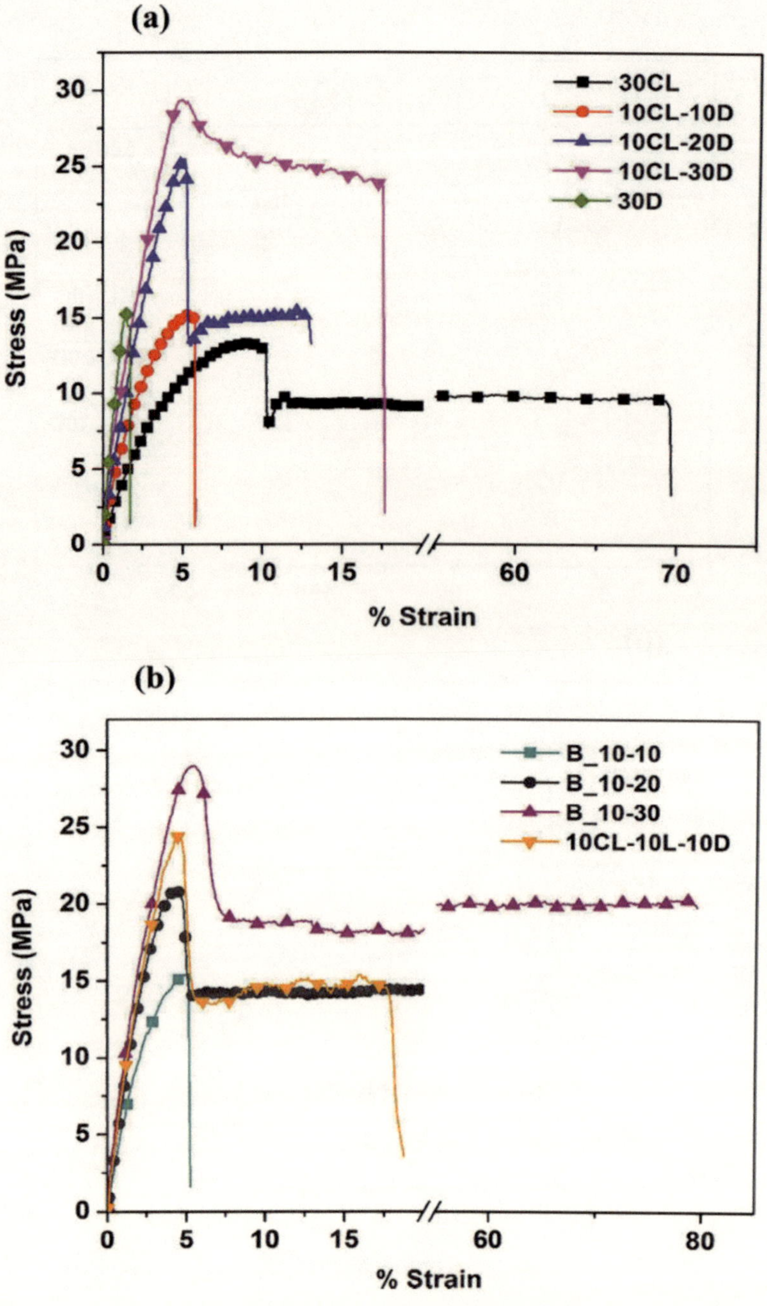

Figure 34.
Representative data for stress vs. (%) Strain of (a) homopolymers and diblock copolymers; (b) diblock blends and trisb copolymer (adapted from reference [12] with a permission).

unusual increase in elongation at break for 10CL-30D, it is important to consider the structure at the amorphous state at higher temperature. Here, it is noted that PCL/PLLA polymer blends exhibit LCST (lower critical solution temperature) phase behavior [44] so that PCL-PLLA blends are subjected to microphase separation at higher temperature. As a matter of fact, the SAXS results (**Figure 35**) indicated the microphase separation at higher temperature (210°C) for the PCL-PDLA diblock copolymers. The judgment of morphology was uncertain due to the presence of only a first order peak. For the 10CL-30D specimen, there may be the

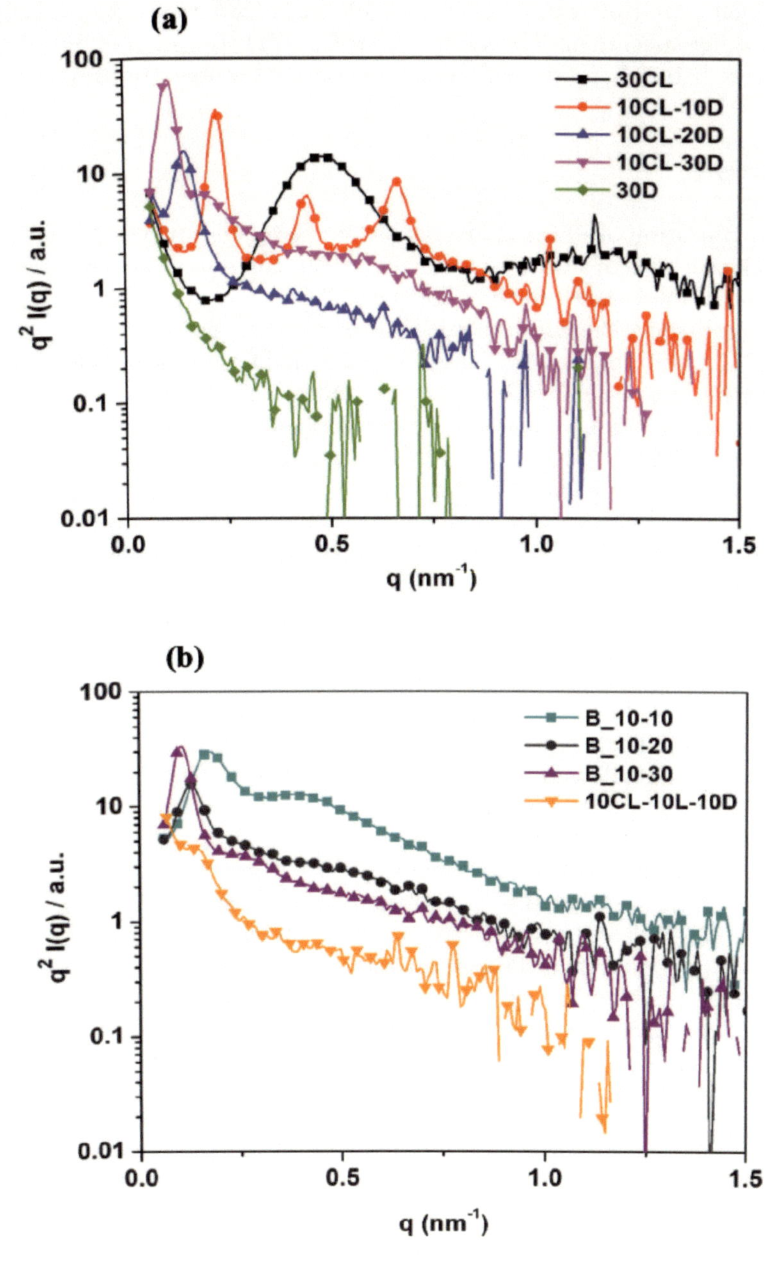

Figure 35.
Lorentz-corrected scattering intensity vs. q of (a) homopolymers and diblock copolymers and (b) enantiomeric diblock copolymer blends and trisb copolymer (adapted from reference [12] with a permission).

possibility of the formation of PCL lamellar or cylindrical microdomains due to the PCL fraction being 25%. It may be perceived that upon rapid quenching from 210 to 0°C, the PDLA matrix is vitrified due to which the PCL block chains would only crystallize in the interior of the cylindrical microdomains surrounded by the glassy matrix. Because of the confined crystallization, the crystallite may be considered as tiny and dispersed in the interior of PCL microdomains. In such a situation, the crack propagation of glassy PDLA matrix is terminated at the PCL microdomains when the specimen is drawn which could be attributed to amorphous PCL phase (rubbery domain) at room temperature. Also, the PCL block chains are much easier to be unfolded from the tiny crystalline lamellae as compared to larger (thicker) lamellae in other specimens. Therefore, the 10CL-30D specimen is found to exhibit the most stretchable character which may be ascribed to its structural origin.

6. Conclusions

The crystallization of PLLA is one of the key factor for analyzing structure–property relationships of PLLA-based blend, block copolymer and nanocomposites. The presence of solid state additives (SFN and CNC) increased the nucleation of PLLA, thus influences the whole crystallization process, however the spherulite growth of PLLA was not significantly changed by loading SFN or CNC. For the case of liquid-state additive i.e. OMG, nucleation and spherulite growth rate both were found to be increased which improves the crystallizability of PLLA. The presence of SFN enhanced the SC crystallization while it suppressed the HC crystallization. It is noteworthy for this particular case that the spherulite growth rate was suppressed by the addition of 1% SFN whereas the nucleation density was much increased by SFN. For the case of PLLA/PEG(50/50) blend, a two-step temperature-jump was conducted as 180.0°C → 127.0°C → 45.0°C. For this particular condition, it was found that PEG can crystallize only in the preformed spherulites of PLLA. The confined crystallization of PEG can be accounted for as follows. Upon the PLLA crystallization at 127.0°C, the PEG content in the amorphous region inside the PLLA spherulite is increased because of the formation of the pure solid phase of PLLA (crystalline phase). Then, T_f of PEG increases so that PEG can crystallize but this crystallization is only allowed inside the PLLA spherulite. The direct evidence of the PEG crystallization was obtained by the bright-field optical microscopic observation and WAXS measurements. For the case of PLA and PCL block copolymers, the poor mechanical property of PDLA (which is poor elongation at break) was reported to be much improved by the presence of extensible PCL block component even in case of a short PCL chain in the block copolymer, which may be ascribed to the confined crystallization of PCL in the microdomain structure. Furthermore, the formation of SC crystal in the enantiomeric blend of PLLA-*b*-PCL and PDLA-*b*-PCL have been examined by changing the block length of PLLA and PDLA components.

Author details

Amit Kumar Pandey[1] and Shinichi Sakurai[1,2*]

1 Department of Biobased Materials Science, Kyoto Institute of Technology, Japan

2 Indian Institute of Technology Guwahati, India

*Address all correspondence to: shin@kit.ac.jp

References

[1] Saeidlou, S.; Huneault, M.A.; Li, H.; Park, C.B. Poly(lactic acid) crystallization. *Prog. Polym. Sci.* **2012**, *37*, 1657–1677.

[2] Wasanasuk, K.; Tashiro, K.; Hanesaka, M.; Ohhara, T.; Kurihara, K.; Kuroki, R.; Tamada, T.; Ozeki, T.; Kanamoto, T. Crystal structure analysis of poly(l-lactic acid) α form on the basis of the 2-dimensional wide-angle synchrotron X-ray and neutron diffraction measurements. *Macromolecules* 2011, *44*, 6441–6452.

[3] Wasanasuk, K.; Tashiro, K. Structural regularization in the crystallization process from the glass or melt of poly(l-lactic acid) viewed from the temperature-dependent and time-resolved measurements of FTIR and wide-angle/small-angle X-ray scatterings. *Macromolecules* **2011**, *44*, 9650–9660.

[4] Zhang, J.; Tashiro, K.; Tsuji, H.; Domb, A.J. Disorder-to-order phase transition and multiple melting behavior of poly(L-lactide) investigated by simultaneous measurements of WAXD and DSC. *Macromolecules* **2008**, *41*, 1352–1357.

[5] Zhang, J.; Tashiro, K.; Domb, A.J.; Tsuji, H. Confirmation of disorder α form of poly(L-lactic acid) by the X-ray fiber pattern and polarized IR/Raman spectra measured for uniaxially-oriented samples. *Macromol. Symp.* **2006**, *242*, 274–278.

[6] Aleman, C.; Lotz, B.; Puiggali, J. Crystal Structure of the α-Form of Poly (L-lactide). *Macromolecules* **2001**, *34*, 4795–4801.

[7] Thi Ngoc Diep, P.; Mochizuki, M.; Doi, M.; Takagi, H.; Shimizu, N.; Igarashi, N.; Sasaki, S.; Sakurai, S. Effects of a special diluent as an agent of improving the crystallizability of poly (L-lactic acid). *Polym. J.* **2019**, *51*, 283–294.

[8] Pandey, A.K.; Katiyar, V.; Sasaki, S.; Sakurai, S. Accelerated crystallization of poly(l-lactic acid) by silk fibroin nanodisc. *Polym. J.* **2019**.

[9] Pandey, A.K.; Katiyar, V.; Takagi, H.; Shimizu, N.; Igarashi, N.; Sasaki, S.; Sakurai, S. Structural Evolution in Isothermal Crystallization Process of Poly(L-lactic acid) Enhanced by Silk Fibroin Nano-Disc. *Materials.* **2019**, *12*, 1872.

[10] Thi Ngoc Diep, P.; Takagi, H.; Shimizu, N.; Igarashi, N.; Sasaki, S.; Sakurai, S. Effects of Loading Amount of Plasticizers on Improved Crystallization of Poly (L-lactic acid). *J. Fiber Sci. Technol.* **2019**, *75*, 99–111.

[11] Pandey, A.K.; Diep, P.T.N.; Patwa, R.; Katiyar, V.; Sasaki, S.; Sakurai, S. DSC and SWAXS Studies on the Effects of Silk Nanocrystals on Crystallization of Poly(l-Lactic Acid). In *Advances in Sustainable Polymers*; Springer, **2020**; pp. 321–339.

[12] Mulchandani, N.; Gupta, A.; Masutani, K.; Kumar, S.; Sakurai, S.; Kimura, Y.; Katiyar, V. Effect of Block Length and Stereocomplexation on the Thermally Processable Poly (ε-caprolactone) and Poly (Lactic acid) Block Copolymers for Biomedical Applications. *ACS Appl. Polym. Mater.* **2019**, *1*, 3354–3365.

[13] BANPEAN, A.; SAKURAI, S. Confined Crystallization of Poly (ethylene glycol) in Spherulites of Poly (L-lactic acid) in a PLLA/PEG Blend. *Polymer.* **2021**, 123370.

[14] Mondal, K.; Sakurai, S.; Okahisa, Y.; Goud, V. V.; Katiyar, V. Effect of cellulose nanocrystals derived from *Dunaliella tertiolecta* marine green algae

residue on crystallization behaviour of poly(lactic acid). Submitted to *Carbohydr. Polym.* **2020**.

[15] Pandey, A.K.; Takagi, H.; Igarashi, N.; Shimizu, N.; Sakurai, S. Exclusive Formation of Stereocomplex Crystallites in Poly(L-lactic acid)/Poly(D-lactic acid) Blends Promoted by Silk Fibroin Nanodisc. Submitted to *Macromolecules* **2021**.

[16] Li, H.; Huneault, M.A. Effect of nucleation and plasticization on the crystallization of poly (lactic acid). *Polymer.* **2007**, *48*, 6855–6866.

[17] Carbone, M.J.; Vanhalle, M.; Goderis, B.; Puyvelde, P. Van Amino acids and poly (amino acids) as nucleating agents for poly (lactic acid). **2015**, *35*, 169–180.

[18] Kawamoto, N.; Sakai, A.; Horikoshi, T.; Urushihara, T.; Tobita, E. Nucleating agent for poly(L-lactic acid) - An optimization of chemical structure of hydrazide compound for advanced nucleation ability. *J. Appl. Polym. Sci.* **2007**, *103*, 198–203.

[19] Ouchiar, S.; Stoclet, G.; Cabaret, C.; Gloaguen, V. Influence of the Filler Nature on the Crystalline Structure of Polylactide-Based Nanocomposites: New Insights into the Nucleating Effect. *Macromolecules* **2016**, *49*, 2782–2790.

[20] Müller, A.J.; Ávila, M.; Saenz, G.; Salazar, J. *CHAPTER 3. Crystallization of PLA-based Materials*; 2015; ISBN 9781782624806.

[21] Patwa, R.; Soundararajan, N.; Mulchandani, N.; Bhasney, S.M.; Shah, M.; Kumar, S.; Kumar, A.; Katiyar, V. Silk nano-discs: A natural material for cancer therapy. *Biopolymers* **2018**, *109*, 23231.

[22] Tsuji, H. Poly(lactic acid) stereocomplexes: A decade of progress. *Adv. Drug Deliv. Rev.* **2016**, *107*, 97–135.

[23] Tan, B.H.; Muiruri, J.K.; Li, Z.; He, C. Recent Progress in Using Stereocomplexation for Enhancement of Thermal and Mechanical Property of Polylactide. *ACS Sustain. Chem. Eng.* **2016**, *4*, 5370–5391.

[24] Tsuji, H.; Ikada, Y. Stereocomplex Formation between Enantiomeric Poly (lactic acid)s. 9. Stereocomplexation from the Melt. *Macromolecules* **1993**, *26*, 6918–6926.

[25] Okihara, T.; Tsuji, M.; Kawaguchi, A.; Katayama, K.I.; Tsuji, H.; Hyon, S. H.; Ikada, Y. Crystal Structure of Stereocomplex of Poly(L-lactide) and Poly(D-lactide). *J. Macromol. Sci. Part B* **1991**, *30*, 119–140.

[26] Tsuji, H.; Fukui, I. Enhanced thermal stability of poly(lactide)s in the melt by enantiomeric polymer blending. *Polymer.* **2003**, *44*, 2891–2896.

[27] Long, H.; Wu, Z.; Dong, Q.; Shen, Y.; Zhou, W.; Luo, Y.; Zhang, C.; Dong, X. Effect of polyethylene glycol on mechanical properties of bamboo fiber-reinforced polylactic acid composites. *J. Appl. Polym. Sci.* **2019**, *136*, 47709.

[28] Liu, H.; Zhang, J. Research progress in toughening modification of poly (lactic acid). *J. Polym. Sci. part B Polym. Phys.* **2011**, *49*, 1051–1083.

[29] Lai, W.-C.; Liau, W.-B.; Lin, T.-T. The effect of end groups of PEG on the crystallization behaviors of binary crystalline polymer blends PEG/PLLA. *Polymer.* **2004**, *45*, 3073–3080.

[30] Li, Y.; Li, X.; Xiang, F.; Huang, T.; Wang, Y.; Wu, J.; Zhou, Z. Crystallization, rheological, and mechanical properties of PLLA/PEG blend with multiwalled carbon nanotubes. *Polym. Adv. Technol.* **2011**, *22*, 1959–1970.

[31] Kulinski, bZ; Piorkowska, E. Crystallization, structure and properties

of plasticized poly (L-lactide). *Polymer.* **2005**, *46*, 10290–10300.

[32] Piorkowska, E.; Kulinski, Z.; Galeski, A.; Masirek, R. Plasticization of semicrystalline poly (L-lactide) with poly (propylene glycol). *Polymer.* **2006**, *47*, 7178–7188.

[33] Zhao, X.; Hu, H.; Wang, X.; Yu, X.; Zhou, W.; Peng, S. Super tough poly (lactic acid) blends: a comprehensive review. *RSC Adv.* **2020**, *10*, 13316–13368.

[34] Tsuji, H.; Ikada, Y. Properties and morphologies of poly (L-lactide): 1. Annealing condition effects on properties and morphologies of poly (L-lactide). *Polymer.* **1995**, *36*, 2709–2716.

[35] Strobl, G.R.; Strobl, G.R. *The physics of polymers*; Springer, 1997; Vol. 2;.

[36] Padden Jr, F.J.; Keith, H.D. Mechanism for lamellar branching in isotactic polypropylene. *J. Appl. Phys.* **1973**, *44*, 1217–1223.

[37] Yamada, K.; Matsumoto, S.; Tagashira, K.; Hikosaka, M. Isotacticity dependence of spherulitic morphology of isotactic polypropylene. *Polymer.* **1998**, *39*, 5327–5333.

[38] Uchida, K.; Mita, K.; Higaki, Y.; Kojio, K.; Takahara, A. Lamellar orientation in isotactic polypropylene thin films: a complement study via grazing incidence X-ray diffraction and surface/cross-sectional imaging. *Polym. J.* **2019**, *51*, 183–188.

[39] Hama, H.; Tashiro, K. Structural changes in isothermal crystallization process of polyoxymethylene investigated by time-resolved FTIR, SAXS and WAXS measurements. *Polymer.* **2003**, *44*, 6973–6988.

[40] Patterson, A.L. The scherrer formula for X-ray particle size determination. *Phys. Rev.* **1939**, *56*, 978–982.

[41] Faucher, J.A.; Koleske, J. V; Santee Jr, E.R.; Stratta, J.J.; Wilson Iii, C.W. Glass transitions of ethylene oxide polymers. *J. Appl. Phys.* **1966**, *37*, 3962–3964.

[42] Huang, P.; Zhu, L.; Guo, Y.; Ge, Q.; Jing, A.J.; Chen, W.Y.; Quirk, R.P.; Cheng, S.Z.D.; Thomas, E.L.; Lotz, B. Confinement size effect on crystal orientation changes of poly (ethylene oxide) blocks in poly (ethylene oxide)-b-polystyrene diblock copolymers. *Macromolecules* **2004**, *37*, 3689–3698.

[43] E.W. Fischer, H.J. Sterzel, and G. Wegner, Investigation of the structure of solution grown crystals of lactide copolymers by means of chemical reactions. *Kolloid-Z.Z. Polym.* **1973**, *251(11)*, 980–990.

[44] Meredith, J.C.; J. Amis, E. LCST phase separation in biodegradable polymer blends: poly (D, L-lactide) and poly (ϵ-caprolactone). *Macromol. Chem. Phys.* **2000**, *201*, 733–739.

Chapter 6

Polymorphism and Supramolecular Isomerism: The Impasse of Coordination Polymers

Francisco Sánchez-Férez and Josefina Pons

Abstract

The chapter presented hereafter, outlines the narrow link between chemistry and crystallography that impelled the identification of *polymorphism* and provided a priceless grounding to understand structure-properties relationship. It was initially conceived for organic substances but actually embraced by metal–organic products, especially in the study of coordination polymers. All of the technologic advances have provided profound insights on the control of crystal structures formation revealing that any applied stimulus over a substance can undergo a structural transformation. This has led to the implementation of several methodologies in the industrial and academic segment shedding light on the source of hitherto, not well understood results.

Keywords: polymorphism, pseudopolymorphism, solvates, transition metals, organic linkers, crystal structures

1. Introduction: historical perspective

The intrinsic awareness in structure-properties relationship of solids was firstly introduced on the study of minerals and inorganic compounds and has been preserved over time [1]. Within this frame, the first step on the polymorphism phenomenon was unknowingly given by Klaproth [2] in 1788 identifying three different crystalline phases of calcium carbonate (calcite, vaterite and aragonite). In 1819, Mitscherlich commenced his research on phosphates and arsenates of potassium (KH_2PO_4 and KH_2AsO_4) and established their complete morphological similarity, noting that *they crystallize in similar forms*. His subsequent results with the corresponding potassium salts ($NH_4H_2PO_4$ and $NH_4H_2AsO_4$) confirmed that observation and led him to pose that *there do exist bodies of dissimilar chemical composition having the same crystalline form*. He not only recovered the work of Whollaston with orthorhombic carbonates and sulfates of barium, strontium and lead, who already noticed this phenomenon in 1812 but also, extended to rhombohedral carbonates of calcium, magnesium, iron and manganese, and to sulfates of iron, copper, zinc, magnesium, nickel and cobalt. Curiously, he also evinced the basis of *seeding* at realizing that *having two substances both able to crystallize in various forms, the presence of one during the crystallization of the other will force the crystallization of the latter in the same form*. Then, he moved to Stockholm with Berzelius with whom he delved deeply into phosphates and arsenates and forged the concept of *isomorphism*. Last but not least, he reported two distinct crystallized forms of sulfur in 1826 demonstrating *a clear case of an element which could be made to crystallize in two different systems of symmetry at will, by merely changing the crystallization conditions.* Therefore, Mitscherlich was identified as the one who took the first step towards the rise of *polymorphism* [3]. Shortly after, in 1824, the contradictory results garnered by Liebig [4] and Wöhler [5] during their research on silver fulminate and silver cyanate triggered the conflict which led them to be colleagues and to pose the following dilemma: *can two compounds with the same composition have different physical properties?*. Their results evinced two Ag salts with the same composition but different physical properties, which did not go unnoticed by Wöhler's master Berzelius, who merged these results with those of Mitscherlich. It is unclear who was the first to conceive the notion of *polymorphism* but this crucial period of 1820–1832 was mainly drawn by Mitscherlich [6–8] who shed light on this phenomenon, even though the concept was still vague. It was during these years when Berzelius [9] proposed the concept of *isomerism* (1831–1832) and it took until 1832 for Wöhler and Liebig [10] to report the first case of *polymorphism* in an organic compound. The awareness of *isomerism* set the beginning of structural chemistry, broadening the knowledge and understanding of organic structures.

Since observation was the essential tool to identify *polymorphism*, this research drastically changed with the accessibility and wide spreading use of the microscope, but it was not until 1839 when Frankenheim [11] introduced the first principles defining *polymorphism* and *a postriori*, Mallard [12] set the structural basis of *polymorphism* in 1876, relating differences in physical properties with different arrangements. One of the most remarkable contribution during this period was the "Rule of steps" or "Law of successive reactions" from Ostwald [13] in 1897. He pointed that during a succession of polymorphic forms, those appeared later are generally more stable. Despite not being considered a rule, it is still valid as a general observation. But the two major queries raised by Buerger and Bloom [14] in 1937 were still unanswered: what causes the formation of different phases of a substance and which factors determine them?.

The narrow link between crystallography and *polymorphism* was forged by Tamman [15] in 1926 and settled with the first polymorphic X-ray crystal structure determination of an organic compound, resorcinol, published by Robertson and Ubbelohde [16] in 1939. Despite this achievement, the next decades passed without a better understanding of *polymorphism,* being underrated until 1965 when McCrone [17] conducted a comprehensive study in which he defined a polymorph as: "*a solid crystalline phase of a given compound resulting from the possibility of at least two different arrangements of the molecules of that compound in the solid state*" and published a review in 1969 about the importance of such phenomenon in the pharmaceutical outlook [18]. Since the introduction of the term *allotrope* by Berzelius in 1841, *Polymorphism* had been taken with *allotropy* on the same meaning. But, it was not until the 1990s, when Sharma [19] and Reinke [20] set the differences between them: *polymorphism* occurs in chemical compounds while *allotropy* occurs in chemical elements. This work was crowned *a posteriori* by Dunitz's [21, 22] crystal description contribution.

In the 1970s, the works of Schmidt [23] and Paul and Curtin [24, 25] grounded the flourish of solid-state chemistry and precede the breakthrough of conceiving polymorphism. They served as inspiration to Bernstein and Desiraju, who laid the foundation for recent supramolecular chemistry. In 1978, Bernstein [26] changed the landscape of *polymorphism* by rationalizing the study of crystal packing forces on molecular conformations of polymorphs and later in 1990, together with Etter [27–30], applied their graph set descriptors and provide guidelines to understand polymorphic transformations [31]. Bernstein compiled most of this historical results in his book "*Polymorphism in Molecular Crystals*", which the authors encourage reading [32]. Likewise, Desiraju achieved substantial progress in this field during his studies of structure-properties relationship of organic solids mainly of pharmaceutical interest, emphasizing and aiming its importance in this industry as reflected in his book "*Crystal Engineering: A Textbook*" [33]. Both channeled their polymorphism vision in terms of supramolecular chemistry.

During 1990s, computational chemistry went hand in hand, achieving methodologies capable to reproduce experimental results and enabling even crystal structure predictions. The first attainment was obtained by using Williams' software [34] that met the main handicap hitherto, the identification of lattice energy minimums. Subsequent years, many computational approaches were developed facing with computer-generated structures for prediction, to the extent that in 1999 a collaborative workshop held at the Cambridge Crystallographic Data Centre (CCDC) [35] brought together the benchmark computational groups of this period to provide an objective assessment of the possibilities of crystal structure prediction. The results gathered in this event were clearly summarized *a posteriori* in a paper published by Lommerse [36]. Further advances on prediction methods accuracy as well as the implementation of Density Functional Theory (*DFT*) and Machine Learning (*ML*) can be found in Spark's review [37].

Despite the basis of polymorphism were already defined at the end of the twentieth century, the increasing advances on X-ray diffraction techniques and crystallization methods afforded the determination and analysis of metal–organic structures, especially *coordination polymers*. *Polymorphism* of metal–organic complexes was still unexplored and these new class of materials, from which polymorphic structures grew exponentially, required classification and awareness. This impasse was encouraged by renowned researchers as Sharma [38], Ciani [39], Rogers [40] and Zaworotko [41]. In this regard, the same Zaworotko was who published a review in 2001 emphasizing the difference between *polymorphism* and *supramolecular isomerism* and underlined the link between them in organic and metal–organic networks [42].

2. Types and borders: from organic to metal–organic

During the early stages of polymorphism, the lack of crystallographic tools conditioned the understanding and therefore, classification of the different polymorphic forms, but the exponential advances in this field provided perhaps too many concurrent data to bring it together and form a unique mindset. It was not until 1965 when McCrone [17] gathered the knowledge hitherto and set a more excluding definition of polymorphism. The inclusion or not of hydrated forms and solvates have been discussed since the rise of polymorphism, but be that as it may, he ventured to propose their exclusion from polymorphic forms and to avoid the use of *pseudopolymorphism* to define them. This assertion was underpinned by Bernstein [32] in 2002, even if actually it is a wide spread term, probably because of its acceptance by the pharmaceutical industry [43] from its regulatory and patentability point of view. Since the aim of this chapter is not about *pseudopolymorphs* we therefore, decided to exclude this term to refer to solvates and hydrates, which can lead to undesirable misunderstandings. In 2012, Desiraju, Karpinski, Thaper and Zaworotko [44] defined *cocrystals* as *solids that are crystalline single-phase materials composed of two or more different molecular and/or ionic compounds generally in a stoichiometric ratio which are neither solvates nor simple salts*. As well as *salts* that are *any of numerous compounds that result from replacement of part or all of the acid hydrogen of an acid by a metal or a radical acting like a metal;*

an ionic or electrovalent crystalline solid. Instead, *solvates* have one component, commonly a solvent, which is liquid by itself and *hydrates* are a particular case of *solvates*, containing water as solvent. A deeper discussion on *pseudopolymorphism* can be found in references [45–47].

Until the beginning of the twenty-first century, the polymorphs' classification was settled from the perspective of crystal packing forces and introduced *conformational* and *packing polymorphism* to sort the examples hitherto. *Coordination polymers* are infinite repeating coordination entities composed of organic molecules serving as linkers and metal ions as nodes [48]. Their main classification is based on dimensionality which is the number of directions in which the array is extended. Therefore, they can be divided into one-, two- or three-dimensional. The rise of this field befell in 1961 when Bailar [49] firstly introduced the term *coordination polymers*. Distinction between polymorphs and other forms as solvates, hydrates or cocrystals in organic structures was well defined, but in the case of metal–organic structures, in particular *coordination polymers,* this classification was not sufficient.

Numerous structures which were not truly polymorphs (solvent molecules were present in the lattice), but neither were solvates, started to rise at the end of the twentieth century. This new domain required the borders to be clarified and established. Yet, these rigid structures were thought to be less disposed to suffer structural variations but some examples were gathered in works of Janiak [50] with Zn(II) poly(pyrazolyl)borates, Ciani [39] with Ag(I) and 4-cyanopyridine, Zaworotko [41, 51] with Co(II) and pyridyl containing linkers, Ripmeester [52] with Cu(II) and a diketone and Rogers [40] with Hg(II) and tetrapyridylporphyrines. Thus, *supramolecular isomerism* was invoked to merge those different types of structures (mainly coordination polymers) assembled from identical building blocks. In this impasse, Zaworotko [42] strongly contributed to shed light upon this ambiguity in his review of 2001. He defined *supramolecular isomerism* as "*...the existence of more than one type of network superstructure for the same molecular building blocks...*" and pointed that polymorphs are a particular case of supramolecular isomers "*...polymorphs can therefore be regarded as being supramolecular isomers of one another but the reverse is not necessarily the case*". This assertion should be clear before classifying metal–organic structures.

Henceforth, *supramolecular isomerism* was divided into *structural* (regarding the formation of different networks *inter alia* ladder, brickwall, 3D frame, herringbone, bilayer or Lincoln logs), *conformational* (relying on the flexibility of the ligands) and a new class named *catenane* (promoted by interpenetration), being polymorphic forms a certain condition of them. A schematic representation highlighting such division is depicted in **Figure 1**. At that time, *structural isomerism* was focused on the particular case of *architectural isomers* [41], understood as variations of the connectivity of the ligands between two structures, sharing the same composition as a result of ligand conformations [53]. In particular, to those cases in which the accommodation of different solvent guest molecules promoted the change in the spatial disposition of the organic linkers.

In this direction, Robin and Fromm [48] in 2006 described *supramolecular isomerism* as the ability of a substance to arrange into one or several network superstructures by different molecular or supramolecular assemblies with the inherent condition that organic linkers and metal ion remain the same and this metal ion retains an equal coordination sphere. They also reported several examples of a new subclass of structural supramolecular isomers named as *ring opening*. Recently, Zhou [54] has provided the specific term *framework isomers* to define supramolecular isomerism in MOFs, in which solvent occluded molecules are trivial.

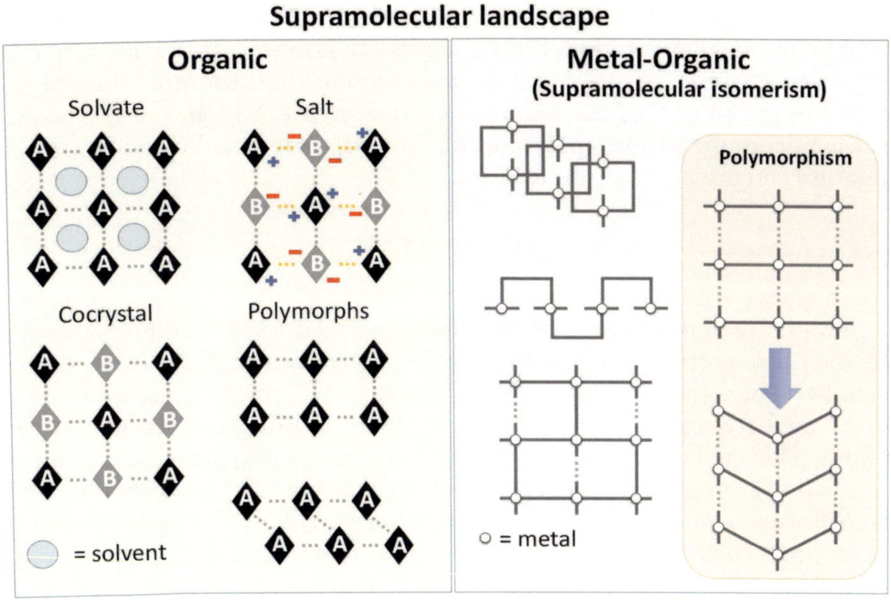

Figure 1.
Schematic representation of the structural relationship between the different architectures among organic and metal–organic compounds.

3. Thermodynamic vs. kinetic factors: formation and transformation of polymorphs

A brief summary of fundamental thermodynamics principles is required prior to understand their relevance on *polymorphism*. Despite ideal systems are not viable, the definition of boundaries within this phenomenon based on simple laws could be achieved with a sufficient accuracy to be useful. It is essential to stress that the result of any structural change will be reflected as a change on the properties of the system. Since energy changes are empirically measurable, they are a pragmatic way to face *polymorphism* and led these energy differences to be a useful descriptor. The identification of the relative intensities within the different forms of energy displays whether exchanges of energy between two polymorphs will occur. Remarkably, this allows to induce energetic modifications, by an external stimulus, only on a selected form of energy to establish relationships in a linear manner.

The quantity of energy a system exchanges with environs is named *enthalphy* and is referred to as H. Taking a chemical reaction as the case study, the character of the energetics involved in it, is represented as the *enthalpy of reaction,* which is the difference in enthalpies between products and reactants. This concept is useful to determine the direction in which energy will flow but such information is not conclusive to establish *spontaneity*. Hence the introduction of the notion of *entropy* (S) is essential. It is a way to represent the organization of a system approached as the degree of disorder and it is measured in terms of entropy changes of the system. Using both terms, *Gibbs free energy* (G) provides further insights into *spontaneity* at constant temperature and pressure.

To understand the relative stability of the different polymorphs within a given system as well as their transformations, the energy/temperature diagrams (E/T) introduced by Buerger in 1951 [55] are the most valuable approach, being capable to encapsulate much data in a single representation. It is based on the G equation:

$$G = H - TS \tag{1}$$

As a general representation, the diagram on **Figure 2a** displays how these terms evolve at increasing temperature. Since the term *TS* is more significantly affected by temperature than *H*, its effect on *G* becomes intense as the temperature of the system raises and thus, *G* of the system tends to decrease. The energy of a transformation undergone within a polymorphic system, under conditions of constant temperature and pressure, are defined by:

$$\Delta G = \Delta H - T\Delta S \quad (2)$$

Within a series of polymorphic forms their crystal lattices are unique *per se*, so despite their *G* at a certain temperature could be equal (*isoenergetic* forms), the values of *H* and *S* will be different (**Figure 2b**). If isoenergetic forms are present, there is a crossing point between the two G value curves in the E/T diagram, named *transition point* (*t.p.*), which to be useful is to be placed below the *melting point* (*m.p.*). In such a scenario, if the formation of these polymorphs is allowed by kinetic factors, both forms coexist and can be formed as *concomitant polymorphs* (*vide infra in Section 5*).

For instance, considering a case study as the one represented in **Figure 2b**, containing two polymorphic forms (*dimorphic*) named as *A* and *B*, a brief analysis of this E/T diagram can provide some useful data. Below the *t.p.* (blue region) form *A* is the most thermodynamically stable as its lower *G* value states. Within this region, its transformation into form B is related to an increase in *H* and it is defined as an *endothermic* transition. At the *t.p.* both forms have the same *G* ($\Delta G = 0$), which applied to Eq. (2) is set that $\Delta H = T\Delta S$, where $\Delta H = H_B - H_A$ and $\Delta S = S_B - S_A$. All these equations allow to quantify the entropic change of a polymorphic transformation once the enthalpic variation ($\Delta H_{t,B\leftrightarrow A}$) is experimentally determined (see Section 4.1). Above the *t.p.* (green region), within this range of temperature and before the change of state at the *m.p.*, form *B* is the thermodynamically stable and therefore, transformation from *A* to *B* is associated with a decrease in *H* (*exothermic* process). After defining these borders, one can infer the proper conditions in which the formation of one polymorph is favored instead of the other.

Considering that theoretical thermodynamic relations applied to systems undergoing phase transformation are bounded by experimental data, more empirical concepts were developed to better represent these changes. Consequently, the terms *Enantiotropism* and *Monotropism* were defined to meet this requirement. The former refers to those systems presenting a reversible transformation before the *m.p*. when heated above or cooled below the *t.p.* Contrarily, *Monotropism* sets that only one polymorph is the more thermodynamically stable over the entire range

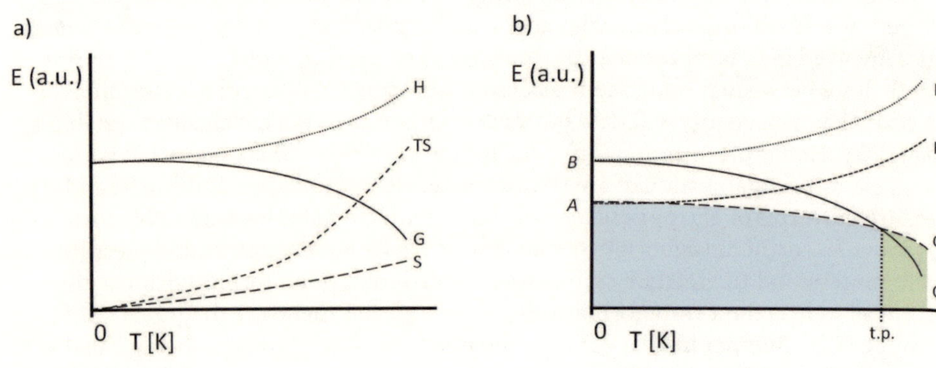

Figure 2.
(a) Progression of the terms G, S, TS and H at increasing temperature. (b) Representation of the evolution of H and G variables in a dimorphic system (polymorphs A and B) at increasing temperature.

of temperatures until the *m.p*. Both examples are illustrated in **Figure 3a** and **b**, respectively.

The incorporation of energetics associated with the liquid state into the E/T diagram, enthalpy (H_{liq}) and free energy (G_{liq}), ease to parse their behavior and classification. If the crossing points between the G_{liq} curve and the G curves of both A and B are after the *t.p.*, they are said to be *enantiotropes*. The absence of crossing points between A and B before their *m.p.* classify them as *monotropes*. The terms $\Delta H_{f,}B$ and $\Delta H_{f,}A$ are the energy that B or A require to change into liquid state, respectively.

Burger and Ramberger [56] in 1979, pioneered in the development of many experimental rules to empirically establish the nature of this relationship between polymorphic forms and their applicability was exemplified with 113 substances [57]. All giving fundamental knowledge to determine the proper conditions to control their formation: Heat-of-transition, Heat-of-fusion, Entropy-of-fusion, Phase transformation reversibility, Enthalpy of sublimation, Heat-capacity, Infrared, Solubility and Density rules. Among them, the most applied are Heat-of-transition and Heat-of-fusion rules. The former being based on using Differential Scanning Calorimetry (DSC) to identify the endothermic or exothermic character of the transition. If this rule cannot be applied, Heat-of-fusion rule, which rely on the determination of the *m.p.*, states that the relation is *enantiotropic* when the higher melting polymorph has the lower enthalpy of fusion while *monotropic* systems are defined by the higher melting polymorph having the higher enthalpy of fusion. Detailed explanation on the application of them could be found in Burger publication [57] or in the more recent Brittain's review [58].

Crystallization is the process related with nucleation and growth of a crystal structure. This process is, in principle, directed by thermodynamic factors that tend to reach the structure with a lower lattice energy. However, the crucial stage is supersaturation which is determined by kinetics, in particular, by the rate of nucleation and thus, the first structure to be formed is the one with preferred nucleation. Therefore, a metastable form can grow despite being unfavored by thermodynamic factors. Subsequently, this form can be converted into a more thermodynamically stable by solution or solid state-mediated phase transformation. This successive phase change was identified and proposed as the *Law of States* by Ostwald [13].

To avoid common mistakes in ascribing the nature of the *nucleation* process, Mullin [59] divided it into *primary*, in which any crystalline matter is directing the process, and *secondary*, pertaining to the circumstances of nucleation generated in the vicinity of crystals, previously nucleated or intentionally placed, a methodology known as *seeding* (*vide infra* in Section 4.2). Furthermore, secondary nucleation can be classified as *homogeneous* if it is spontaneous or *heterogeneous* if it is induced by foreign particles.

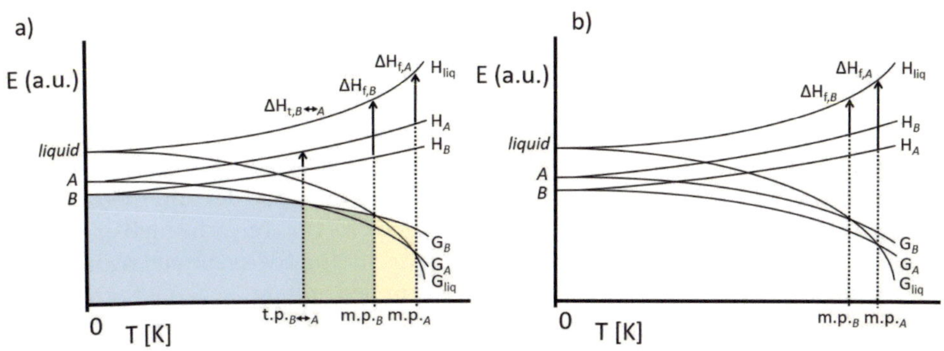

Figure 3.
E/T diagrams of (a) enantiotropic and (b) monotropic dimorphic systems.

Solid-state phase transformations are usually promoted by an external stimulus in the form of mechanical work or temperature. They are thought to be related with intrinsic defects, whether coming from the original structure or being caused by mechanical stress, which commence and propagate the formation of the new phase. The most studied transformations are those based on order–disorder changes, mainly promoted by temperature variations. *Disordering* processes arise from increasing temperature whereas *ordering* processed are observed at decreasing temperature. An excellent analysis of such phenomena with the most relevant examples were summarized by Dunitz and Bernstein [60].

The reversibility of these changes could not be evident or even accessible since *hysteresis* plays a crucial role. It is the lagging of the transition behavior respect to the applied stimulus. Therefore, it is possible to need heating or cooling beyond the *t.p.* to let the phase change occurs and even with the sufficient high degree of *hysteresis* is conceivable to avoid transformation [61]. Although ideal reversibility is often desired, control of hysteresis [62] also leads to unique properties [63], often advantageous to the application of these materials, only achievable by a precedent transformation. *Hysteresis* phenomenon is associated to *structural fatigue* [64–66], which means the rise of structural changes in the crystalline material as dislocations as well as to a different nucleation with different energy barriers from one to another [67]. Further concerns regarding the basis of this topic can be addressed reading Flanagan's publication [68].

Overall, since crystallization is a competitive process between minimizing lattice energy (thermodynamic) and reaching supersaturation (kinetic), the achievement of non-minimal energetically stable forms allows to the transformation into lower energetic forms after reaching the activation barrier.

4. Strategies to identify and achieve polymorphs and the influence on their properties

4.1 Identification of polymorphs

The identification of polymorphism had been, in the vast majority of cases, a matter of chance, but this all changed as a result of the efforts of many scientists in the field. The first tool of recognition was visual observation inasmuch as it was the most affordable and accessible technique. Since it is not a strict approach, however, it is often useful to detect anomalies. In particular, observation of crystalline materials through optical microscopy can allow to differentiate between two crystal *habits* (their characteristic external shape), because this shape is governed by its molecular packing and intermolecular interactions (internal structure). Notwithstanding that more than a century has passed, it is still a routinely used technique, even if optical microscopy is being superseded by Scanning Electron Microscopy (SEM), which allows more accurate surface topology and morphology analysis. The closely related Transmission Electron Microscopy (TEM) was not appropriate for this kind of materials even given its uniqueness. It could collect structural and dynamic data from single crystals in a bulk powder and therefore, have remarkable benefits compared to any other. Unfortunately, the high energy applied to the samples caused strong damages precluding their characterization. Over years, researchers intended to reduce the applied voltage, seeking for this technique to be applicable by minimizing sample damage and improving cameras and detector technologies. This was especially aimed to the study of metal organic frameworks (MOFs), one of the most emerging topics with fundamental need to understand structure-properties relationship. Hence, breakthrough developments in characterization techniques as cryogenic-TEM (cryo-TEM) [69] and High-Resolution TEM (HR-TEM) [70] paved the way to structural features at the nanometric scale as well as recent improvements for *in situ* measurements *inter alia* hot stage TEM, liquid cell TEM (LCTEM) [71] or environmental TEM (ETEM) [72] enabled to observe the dynamics of these systems. Advanced 3-dimensional electronic techniques as automated electron diffraction tomography (ADT) [73] and Rotation Electron Diffraction (RED) [74] were able to gather sufficient data for *ab initio* structure elucidation and thus succeeded where conventional diffraction techniques failed [75].

Despite all these advances, the most valuable technique has been and remains, single-crystal X-ray diffraction (SC-XRD). Only neutron diffraction is tantamount, being capable to collect specific and accurate data of atomic positions, bond distances, and angles [76]. Thus, serving as a complementary technique to SC-XRD. However, one must consider the possibility of temperature driven single-crystal to single-crystal transformations since a growing number of examples have been reported in the literature. Careful inspection of temperature effects on the sample is to be required. The generally used condition in SC-XRD is about 100 K which can undergo the phase transformation. If such events are not assured, one could fail at drawing conclusions of property changes from structural differences.

Although SC-XRD provides complete information about atom positions and structural packing, the growth of suitable crystals for structure determination is sometimes a laborious and very time-consuming task or even not attainable. Often, but nowadays less and less, it is not thought to be part of the endeavors of a chemist. That is why structural studies goes hand in hand with Powder X-Ray diffraction (PXRD), which is in many cases more available and can reflect any structural difference between SC-XRD and the bulk powder. But one must not forget that after ensuring no phase transformation, SC-XRD and neutron diffraction are the unique unambiguous techniques while the rest requires to be combined to successfully identify *polymorphism*. Recent advances in diffraction methodologies have enabled to improve PXRD characterization. For instance, variable temperature-PXRD and variable temperature-SC-XRD not only ease to determine differences in crystalline materials but also allow to trace phase transformations being subject to temperature changes.

Solid-state spectroscopic techniques are also a complementary tool during the identification of structural differences. Sometimes these changes are not evident but, in many cases, subtle structure modifications are reflected in the spectra. The most marked differences observable by Fourier Transformation Infrared Spectroscopy-Attenuated Total Reflectance (FTIR-ATR) [77] or Raman spectroscopies [78] usually appears in the fingerprint region since it is unique for a substance. In the case of polar molecules, transitions associated with rotation can be measured in absorption or emission by microwave or far infrared spectroscopies [79]. Also solid-state Nuclear Magnetic Resonance (SS-NMR) [80] as well as solid-state Ultraviolet–Visible absorption (SS-UV–Vis) [81] and fluorescence [82] have proven to be fruitful techniques to identify polymorphism and phase transformations. The recording of SS-NMR data can be improved by using Schaefer and Stejskal [83] experiments, in which high power heteronuclear decoupling, cross polarization (CP) and magic-angle spinning (MAS) are combined. Careful attention should be paid during MAS since the required high spinning rates generate mechanical stress and local heating, thus favoring conditions for transformations. *Isotopomeric polymorphs* [84] have also been identified, albeit in a lesser extent.

Solid-state Electron spin resonance (ESR) also known as Electron Paramagnetic Resonance (EPR) can be used only for materials containing paramagnetic metal ions or structures in which those metal ions have been embedded. Copper(II) but also cobalt(II) are the archetypal metals for this technique and there already exist examples incorporating Cu(II) into the structures of templated materials [85]. More sophisticated variations include variable-temperature magnetic-susceptibility and variable-temperature solid-state EPR measurements. An increasingly common strategy combines them in the study of single-ion magnets (SIMs), a type of single-molecule magnets (SMMs). This EPR analysis is not easily available since it is preferably implemented with a synchrotron radiation source [86]. The magnetic evaluation of SIMs is performed in solid state so the structural differences between polymorphic forms promoted by conformational changes could lead to a dramatic alteration of the magnetic properties [87]. This effect is marked in Clathrochelates, a special class of structurally rigid cage metal complexes [88].

Last but not least, thermoanalytical techniques *inter alia* hot stage microscopy (also known as thermal microscopy), Thermogravimetric analysis (TGA), Differential Thermal Analysis (DTA) and Differential Scanning Calorimetry (DSC), are widely used to characterize both polymorphism as well as phase transformation phenomena. They are usually combined to maximize the efficiency of the data collection as TG-DTA or DSC-TGA, among others.

During hot stage microscopy, the sample is subjected to heating and cooling processes under polarized light. It provides *m.p.* data as well as, if it occurs, the reversible or irreversible character of the transition. TGA method is used to determine the thermal stability of the products and as previously mentioned, it is commonly combined with DTA. By the use of a thermobalance, it is capable to measure weight losses during temperature changes. It is especially useful to determine desolvation temperatures and thermal stability ranges but it is less accurate to quantify these transitions. The most appropriate technique to track phase transformations with quantitative data is DSC.

DSC is routinely used to measure the difference in the amount of heat required to increase the temperature of the sample respect to a reference. It is divided into power compensation DSC and heat-flux DSC methods. The quantification of the $\Delta H_{t,B\leftrightarrow A}$ enables to identify its exo- or endothermic character, as well as the determination of reversible or not transitions provides essential data about the *enantiotropic* or *monotropic* behavior. DSC has not been deeply exploited yet for metal–organic materials but this practice is inevitably set to change. In particular, it is increasingly used in the study of breathing metal–organic frameworks (MOFs) [89] or solid-state phase transformations in Zeolitic imidazolate frameworks (ZIFs) [90].

Evidence of *polymorphism* can also be confirmed by nanoindentation or by optical properties as refractive index (n) or the identification of an interference figure caused by birefringence, which is the presence of different n and mainly depends on crystallographic orientations [91]. Once *polymorphism* has been identified and characterized, the proper conditions to isolate or to only reach one polymorphic form are to be established, avoiding the presence of mixtures and undesired products.

4.2 Screening and isolation of polymorphic forms

There are significant factors determining the formation of polymorphs *inter alia* molecular structure, chemical composition, energetic differences and experimental conditions (solvent, additives, pH, temperature and pressure). What should be clear is that the different polymorphic forms of a given structure can be selectively reach either by crystallization from the melt or solution, or by solid-state transformation. Crystallization approach has been the most widely studied heretofore but awareness of hitherto ignored solid-state transformations has led to value their tantamount importance. The solution-mediated approach is based on the proper adjustment of the crystallization process and involves much more control of the polymorphs forming conditions. Instead, the advances in solid-state characterization techniques have triggered a significant increase of polymorphism studies allowing traceability of such conversions. This is especially the case of *coordination polymers* and MOFs, optimal materials for polymorphism study because of their flexibility and capability to accommodate structural modifications without the breaking of bonds.

The first step in polymorphs screening is to determine the phase space of a substance and the boundaries of stability for the different forms as well as identifying, if it is the case, interconversion. Defining the most stable phase is recommended since, unless modulating external factors dictate otherwise, that form would be the result. The occurrence of polymorphs and their transformations are confined to what is known as *occurrence domains* that encompass all the conditions in which the targeted crystal forms originate. Early studies carried out by Sato [92] on stearic acid delved deeply into the dependence of temperature and supersaturation on solvent polarity. Is in those regions with *domain* overlap where polymorphic transformations can occur, bearing in mind that the *domain* is not unique for one crystallographic form.

The many attempts to control the formation of a desired form have supplied us of a vast number of methods to selectively achieve it through crystallization or solid-state phase transformation. Most of the old and recent methods have been compiled in **Figure 4**. Further details about fundamental crystallization methods are found in Hulliger's review [93].

Over all of the difficulties of achieving isolated polymorphs, to identify the conditions to reach isolated forms is an essential task. When dealing with a polymorphic mixture scenario, the initial way of facing it, is the use of common crystallization methods as those mentioned before. However, crystallization of less stable forms is often intricated and therefore, it requires the design of more robust strategies *inter alia* high-pressure crystallization, spray-drying, crystallization from a melt or crystallization from a quenched amorphous phase. They give sometimes

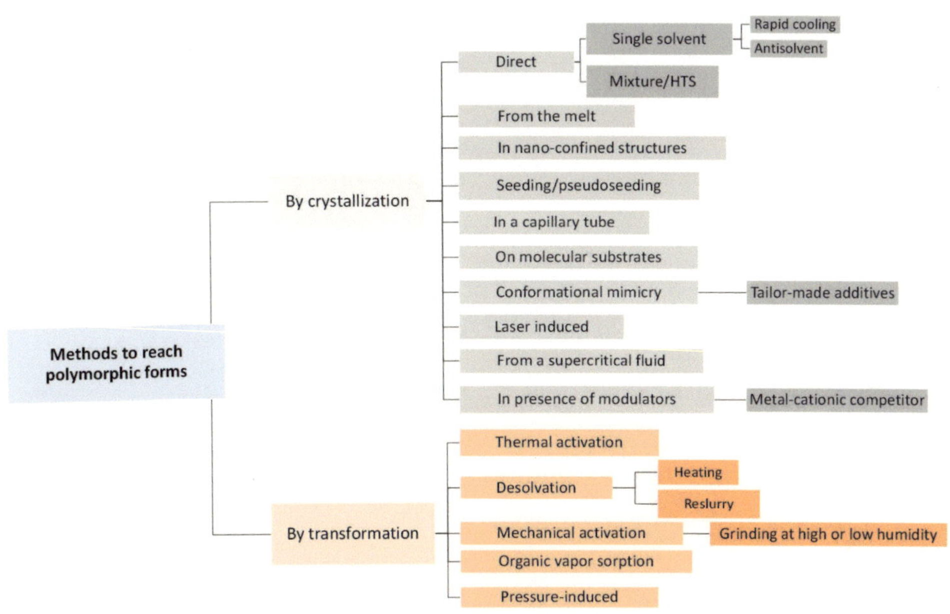

Figure 4.
Classification of the different methods to achieve polymorphic forms by crystallization or transformation.

satisfactory results, but their major drawback is the lack of control in the formation of a single product. Hence, the use of additives and substrates was implemented as template though a limited triumph, considering that only thermodynamic aspects are contemplated and kinetic factors have a determinant role in nucleation. Subsequent methods as application of external fields, surface templating, selective nucleation by supersaturation control and nucleation temperature or seeding experiments emerged, but there is still a need for their improvement.

It is nevertheless important to note again that MOF materials are themselves appropriate candidates for filing structural modifications, since the predefined preferences of the organic linkers combined with those of the metal ions result in a restricted range of potential structures. This is strongly reflected in the common formation of isostructural products although they combine different linkers and metal ions. Such a controllable way of structure design is therefore adequate to identify new strategies for the isolation of different crystalline phases. Currently, throughout all the advances in MOFs design, *polymorphism* and *isomerism* awareness has driven the seeking of selective crystalline phase formation methods. For instance, in the case of mixed-metal MOFs, in which the addition of more than one metal ion disrupts the predefined structural formation, the selective phase formation has been achieved by the incorporation of guiding organic linkers or even metal ions to template the structural assembly. There was already consciousness of *polymorphism* and *isomorphism* in chromium(III) terephthalate MOFs [94], but

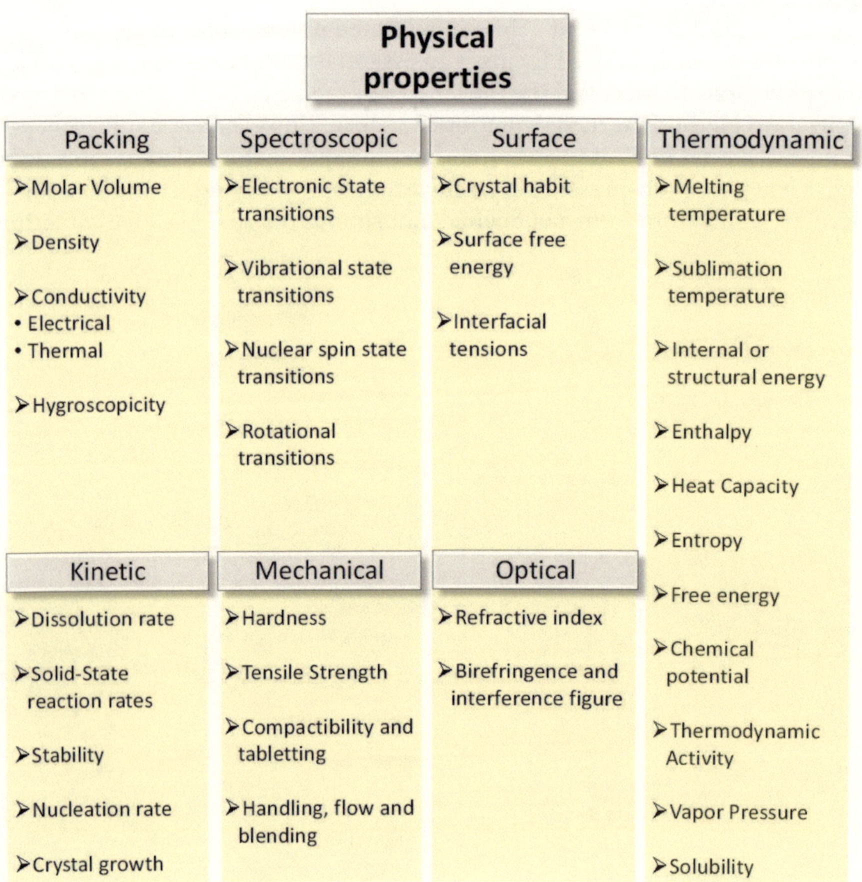

Figure 5.
Properties likely to be altered by polymorphic modifications.

the strategies to reach them were lacking. In 2018, Bureekaew [95] controlled the formation of this family of MOFs by using iron(III) metal-cationic competition, which served as *modulator* during crystallization. Likewise, Užarevic [85] demonstrated a rapid and selective way of controlling *polymorphism* in this family of MOFs by a mechanochemical approach using additives.

4.3 Properties

Become aware of a difference in the properties of a unique sample or two sample which in principle, are to be identical is the commencement of most *polymorphism* studies. One important aspect to be considered, sometimes careless, is the manipulation of the samples before the measurements. It is widely reported that a routinely sample treatment as grinding can undergone phase transformation. Even this subtle stimulus can provoke changes in the sample with the consequent incongruent results, thus hampering their correct analysis. These changes produced by polymorphic modifications could range from slight differences to dramatic alterations of the target properties and those which have raised more interest are listed in **Figure 5**.

5. The special cases: concomitant, vanishing (or disappearing) and intergrowth polymorphism

The terms *Vanishing or disappearing polymorphs* were renowned during the second half of the twentieth century by referring to the evolving nature of some crystal forms over time that caused their unrepeatability. In the course of the subsequent years, researchers realized that every crystal form can be repeated being the finding of the proper conditions (control over nucleation and crystal growth) the main trouble.

Within the polymorphism landscape, the presence of mixtures of crystalline phases has brought a profound attention whereby, many efforts were devoted to the understanding of such circumstances. Reproducibility and purity have always been a requisite for chemists, so such demonstration of lack of control led them to seek to comprehend and identify the source of these phenomena. These two closely related special cases, named as *Concomitant* and *Vanishing* polymorphs, refers to a condition describing all the forms involved. *Concomitant* polymorphs are those simultaneously formed at the exactly same conditions. This assertion is not trivial since many factors contribute to the crystallization process. The case of *vanishing* or *disappearing* polymorphs describes the formation of a metastable phase which undergoes transformation into a more thermodynamically stable one. This phenomenon is quite thorny since as mentioned, reproducibility of the results is often intricated. Such conversions can be found by redissolution of the former crystal or in solid-state. In the solution-mediated process, the formation and disappearing of a metastable product can be quenched by *seeding*. A crucial statement is to be noted, once seeds of one polymorph are formed, the other form will no longer be formed. By using this criterion, seeding *crystallites* of the desired polymorph will lead to the growth of its crystals, even though it is a metastable form [60]. This method allows to avoid both undesirable cases, *concomitancy* and *vanishing*, if phase-pure materials are to be achieved. *Vanishing* by solid-state transformations is even less evident, indeed, any slight appliance of an external stimuli can promote this phase change and routinely sample treatments required for many characterization techniques as grinding, milling or pressure and temperature changes could be sufficient to trigger it. Both phenomena have probably been less reported than occurs and just a curious point is that the first recognized example of *polymorphism* in an organic substance described by Wöhler [10] the *dimorphism* of benzamide, that was also the first precedent of *concomitancy*.

The critical point describing *concomitant* and *vanishing* polymorphs is based on the same kinetic/thermodynamic factors, better understandable by the mentioned E/T plots and promoted by the kinetic govern. In fact, *concomitancy* is dependent on where from the diagram the polymorphs are growing. Having this data, one can also design a strategy to favor the nucleation of one polymorphic form instead of the other.

Identification of concomitant polymorphs is as always, initially assessed by visual recognition. Thus far, careful inspection during hot-stage microscopy has been the most reliable method. Differences in crystal habit and variable melting point suggesting crystalline mixtures could provide a clue as to trace *concomitancy*. Also broadening signals in solid-state NMR or FTIR-ATR may be a symptom of this.

Reported cases of *vanishing* polymorphs in metal–organic compounds are in large part still unfathomed. The phenomenon of *concomitant* polymorphs is equally meager being only few examples reported hitherto. Oliver [96] in 2012 reported the special case of two Cd(II) coordination polymers concomitantly formed from which the less stable form was subsequently identified as a disappearing polymorph. Both products crystallize in the monoclinic crystal system but the stable form (2) exhibited a C_2/c space group and bigger unit cell parameters than the vanishing form (1), which displayed a P2/c space group. Their main dissimilarity was the slightly different orientation of the dipicolinate ligands (**Figure 6**), which improved the inter-chain π···π interactions in the structure of the stable form and provoked a different packing.

In the case of organic structures, several examples can be found in literature. Chalcones are a class of natural products widely used in medicinal chemistry. For instance, the (E)-3′-dimethylamino-nitrochalcone has demonstrated *concomitant polymorphism*, easily detectable by the different colors associated to each form [97].

In spite of topologically flexible MOFs as ZIFs and other azolates are prone to manifest *polymorphism*, the scarce monitoring of *in situ* structure formation thwart its recognition. An example of concomitant polymorphism has been recently reported by Sánchez and Fernández [98] with two Pt(II) metallosupramolecular polymers. Both products were formed by self-assembly of monomeric units but

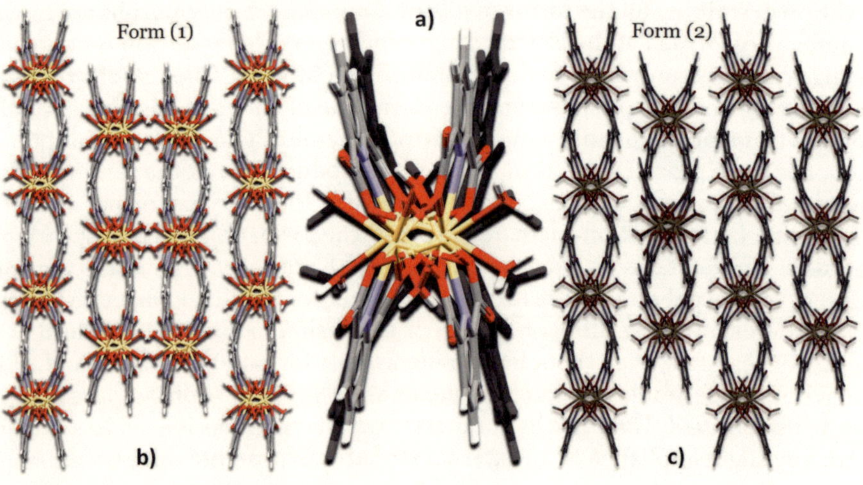

Figure 6.
(a) Overlapping representation of the two Cd(II) polymeric chains in the two polymorphs. Different disposition of dipicolinate ligands: front chain (1) and back chain in dark (2). (b) c axis view of the packing of the less stable form 1 and (c) a axis view of the packing of the more stable form 2 [96].

differed in presenting slipped or pseudoparallel packings. In 2020, a new study from Hanusa and Friščić [99] identified the presence of a *disappearing* polymorph during the formation of two different Hg(II) imidazolate ($Hg(Im)_2$) phases, synthesizing a new layered structure (*sql*) with the consequent disappearing of a previously reported interpenetrated dense phase (*dia*) $Hg(Im)_2$ [100]. Both forms exhibited an orthorhombic crystal system but having evident structural differences driven by an *agostic* interaction (C-H··Hg) in the *sql* form. The *dia*-$Hg(Im)_2$ contained tetrahedral Hg(II) nodes in a Pbca space group and cell parameters of a = 14.5899(3) Å; b = 10.8076(2) Å; c = 9.8200(2) Å while the *sql*-$Hg(Im)_2$ form presented a tetrahedral *see-saw* geometry in the space group $P2_12_12$, with a = 9.4089(4) Å, b = 7.6414(3) Å, c = 5.3625(2) Å. The transient nature of the *dia* form was tracked by PXRD during the mechanochemical synthesis of *sql* form, being inaccessible to reach *dia* form as a final product.

Intergrowth polymorphism was firstly reported by Bond, Boese and Desiraju [101] in 2007 during a study about the doubtful crystalline forms of aspirin and the related difficulties of its structural refinement. It was conceived *to refer to the existence of distinct structural domains within a single crystal of a compound*. When analyzing the one-dimensionally diffuse diffraction data, using Bürgi [102] method, they noticed some diffuse streaks between the Bragg reflections. Considering the reported results of Bürgi, this would be associated with the presence of a less ordered domain. Careful inspection of such results led them to identify two differently ordered domains in the same crystal of aspirin. They also demonstrated by a nanoindentation study in 2014, a bimodal mechanical response depending on which of the crystal faces were measured [103]. From the metal–organic perspective, there had already been examples reported by Ciani of *coordination polymers* of cobalt(II) *intergrowth supramolecular isomers* [104] and copper (II) *intergrowth polymorphs* [105] due to the presence of conformational non-rigid linkers.

6. Interest of polymorphism in organic and metal–organic structures

Polymorphism has an implicit interest since it represents a special situation for the study of structure-properties relationship with limited number of variables as well as provides essential information to understand and control the crystalline inception. The special case of *concomitant polymorphs*, is in turn, an even more worthwhile situation. From a unique reaction, one could establish direct relations between structure modifications and properties. They are also benchmark products for computational analysis as well as for verification of structure-prediction softwares. All this beneficial knowledge can be fruitfully employed throughout the industrial and academic landscape.

6.1 Industrial interest

In the industrial field, stability and purity are mandatory equal for organic and metal–organic materials. Properties of organic solids reflected in their processability and storage as well as solubility and dissolution rates are directly related with bioavailability of a drug and hence are of prime interest to the pharmaceutical industry. For instance, an important concern in the manufacturing, storage and transport of energetic materials is that polymorphic modifications alter the energetics and safety risks associated to them. A representative case is 1,3,5,7-tetranitro-1,3,5,7-tetrazocane (HMX), in which higher compact crystal packings lead to the safer form [106]. In addition, *polymorphism* has relevant commercial impact in the patenting of drugs [107] as in the renowned case of Ranitidine, a drug for the treatment of stomach and intestines ulcers. The subsequent discovery of the polymorphic form 2, having an easier manufacturing procedure, after the patenting of form 1, originated a lengthy court process [108]. *Concomitant* polymorphs are regularly found in industrial precipitation processes in the pharmaceutical and fine-chemicals sectors as in the case of L-histidine, to which anti-solvent crystallization became an inevitable proceeding [109]. *Coordination polymers* has special interest for waste water treatment, protective coatings and fluorescent chemosensors [110]. For instance, in MOFs the thermo-mechanical stability is crucial to move towards the industrial segment. Among the most promising applications, those of industrial interest are adsorption, separation, purification and catalysis. They are being exhaustively tested to supersede, by improved performance, the extensively exploited zeolitic molecular sieves, activated carbon and base metal oxides [111]. Furthermore, ZIFs materials have exhibited remarkable efficiency in separation of olefin/paraffin mixtures [112] and emerged as appropriate candidates to adsorb and retain radioactive iodine [113]. All of these applications depend on the structural arrangement of the materials and thus, control over crystal structure formation is imperative.

6.2 Academic interest

The interest of *polymorphism* in *coordination polymers* lies on the always present structure-properties relationship. The exact control of their structural arrangement is reflected in the achievement of the desired chemical and physical properties. Despite being known as promising functional materials their modular nature can result in polymorphic forms and thus, hampering their application. This is emphasized in applications demanding a high selectivity as enantiomeric separation, gas storage, sensing, molecular recognition, ionic exchange [114], heterogeneous catalysis [115] or non-linear optics. The rise of MOFs, a remarkable case of ordered coordination polymers with potential voids and permanent porosity was driven by the breakthrough of the archetypal MOF-5 [116]. Their main attribute is the controlled porosity to which the formation of cages allows their controlled use in such applications. Currently research is devoted to study structural transformations in Zr-based MOFs. Also a particular family of MOFs essentially constructed with zinc(II) or cobalt(II) metal ions, the Zeolitic Imidazolate Frameworks (ZIFs), have been explored for their superior thermal and chemical stability but as zeolites, the ZIF family displays rich polymorphism [117]. The $Zn(Im)_2$ itself can accommodate 18 polymorphic forms, being essential towards its application the finding of controlled synthesis as the recently established template-mediated route [118]. As afore mentioned, there has been evidence in the recent cases of Cd(II) *coordination polymers* or especially in the ZIF material *dia*-$Hg(Im)_2$ of the undesirable scenario of *vanishing polymorphs.* In the latter case, the formation of the more stable form has hampered the obtention of the 3D structure being superseded by the new layered *sql*-$Hg(Im)_2$. These examples demonstrate the dormant resemblance of metal–organic materials with such well known phenomenon in organic compounds. Another important subject are phthalocyanines, being copper phthalocyanine the model compound. They revolutionized color printing offering a better economic remedy but facing polymorphic troubles since their discovery. Subtle alterations of their crystal packing which is based on $\pi\cdots\pi$ interactions, acutely influences the absorption properties with the consequent color change from blue to red [119]. *Intergrowth polymorphism* and *intergrowth supramolecular isomerism* in coordination polymers has also been promoted by flexible linkers and several examples have been reported hitherto mainly with d^{10} metal ions [120]. This ability to present different properties available in a single crystal open new possibilities for materials design.

7. Conclusions

The perspective of *polymorphism* was primarily based on organic substances until the beginning of the twenty-first century. The increasing advances on solid state characterization, especially *in situ* measurements, benefited the rise of metal–organic structure and its awareness on polymorphic modification. Currently, the ease of tracking structural transformations of dynamic materials has brought to promising insights into the understanding and control of *polymorphism*.

Acknowledgements

J.P. acknowledges financial support from the CB615921 project, the CB616406 project from "Fundació La Caixa" and the 2017SGR1687 project from the Generalitat de Catalunya. F.S.F. acknowledges the PIF pre-doctoral fellowship from the Universitat Autònoma de Barcelona.

Author details

Francisco Sánchez-Férez and Josefina Pons*
Chemistry Department, Autonomous University of Barcelona, Barcelona, Spain

*Address all correspondence to: josefina.pons@uab.es

References

[1] Wells AF. Structural inorganic chemistry. Fifth Edi. Nature. Oxford, UK: Clarendon Press; 1989. p. 294-315

[2] M.H. K. Phosphorsäure, ein Bestandtheil des Apatits. Arch f Miner Geognosie u Hüttenkd. 1798;1:294-300.

[3] Tutton AEH. The work of Eilhardt Mitscherlich and his discovery of isomorphism. In: The natural History of Crystals. First Edit. London, UK: Kegan Paul; 1924. p. 70-97.

[4] Gay-Lussac JL, Liebig J. Analyse du Fulminate d'Argent. Ann Chim Phys. 1824;25:285-311.

[5] Wöhler F. Poggendorffs Ann. Phys. Chem. 1824;1:117-24.

[6] Mitscherlich E. Sur la relations qui existe entre la forme cristalline et les proportion chimiques. Ann Chim Phys. 1820;14:172-90.

[7] Mitscherlich E. Sur la relation qui existe entre la forme cristalline et les proportions chimiques, I. Mémóire sure les arseniates et les phosphates. Ann Chim Phys. 1822;19:350-419.

[8] Mitscherlich E. Über die Körper, welche in zwei verschiedenen krystallisieren Formen. [Considering the materials which can crystallize in two different crystal forms]. Abhl Akad Berlin. 1823;43-8.

[9] Ihde JI. The Development of Modern Chemistry. New York, USA: Dover Publications Inc.; 1984. p. 170-173.

[10] Wöhler F, Liebig J. Untersuchungen über das Radikal der Benzoesäure [Investigations of the radical of benzoic acid]. Ann der Pharm. 1832;3:249-82.

[11] Frankenheim ML. Ueber die Isomerie. Prakt Chem. 1839;16:1-14.

[12] Mallard E. Explication des phenomes optiques anomaux. Ann Mines. 1876;10:60-196.

[13] Ostwald W. Studien über die Bildung und Umwandlung fester Körper 1. Abhandlung: Übersättigung und Überkaltung. Zeitschrift für Phys Chemie Int J Res Phys Chem Chem Phys. 1897;22:289-330.

[14] Buerger MJ, Bloom MC. Crystal Polymorphism. Zeitschrift für Krist - Cryst Mater. 1937;96:182-200.

[15] Tammann GHJA. The states of aggregation: The changes in the state of matter in their dependence upon pressure and temperature (translate by F.F. Mehl). First Edit. New York, USA: Van Nostrand; 1926. p. 116-157.

[16] Robertson JM, Ubbelohde AR. Structure and thermal properties associated with some hydrogen bonds in crystals, II. Thermal expansion. Proc R Soc London Ser A Math Phys Sci. 1939;170:241-51.

[17] McCrone WC. Polymorphism. In: Fox D, Labes M., Weissberger A, editors. Physics and Chemistry of the Organic Solid State. Volume 2. New York, USA: Wiley Interscience; 1965. p. 725-67.

[18] Haleblian J, McCrone W. Pharmaceutical applications of polymorphism. J Pharm Sci. 1969;58:911-29.

[19] Sharma BD. Allotropes and polymorphs. J Chem Educ. 1987;64:404-7.

[20] Reinke H, Dehne H, Hans M. A discussion of the term "Polymorphism." J Chem Educ. 1993;70:101.

[21] Dunitz JD. Phase transitions in molecular crystals from a chemical

viewpoint. Pure Appl Chem. 1991;63:177-85.

[22] Dunitz JD. Thoughts on crystals as supermolecules. In: The crystal as a supramolecular entity Perspectives in supramolecular chemistry. Desiraju GR, editor. Chichester, UK: Wiley; 1996. p. 1-30.

[23] Schmidt GMJ. Photodimerization in the solid state. Pure Appl Chem. 1971;27:647-78.

[24] Paul IC, Curtin DY. Thermally Induced Organic Reactions in the Solid State. Acc Chem Res. 1973;6:217-25.

[25] Paul IC, Curtin DY. Reactions of Organic Crystals with Gases. 1975;187:19-26.

[26] Bernstein J, Hagler AT. Conformational Polymorphism. The Influence of Crystal Structure on Molecular Conformation. J Am Chem Soc. 1978;100:673-81.

[27] Etter MC. Encoding and Decoding Hydrogen-Bond Patterns of Organic Compounds. Acc Chem Res. 1990;23:120-6.

[28] Etter MC, MacDonald JC, Bernstein J. Graph-set analysis of hydrogen-bond patterns in organic crystals. Acta Crystallogr Sect B. 1990;46:256-62.

[29] Etter MC. Hydrogen bonds as design elements in organic chemistry. J Phys Chem. 1991;95:4601-10.

[30] Bernstein J, Etter MC, Macdonald JC. Decoding Hydrogen-bond Patterns. The Case of Iminodiacetic Acid. J Chem Soc, Perkin Trans 2. 1991;1989-91.

[31] Bernstein J, Davis RE, Shimoni L, Chang N -L. Patterns in Hydrogen Bonding: Functionality and Graph Set Analysis in Crystals. Angew Chemie Int Ed English. 1995;34:1555-73.

[32] Bernstein J. Polymorphism in Molecular Cystals. New York, USA: Oxford University Press Inc.; 2002. p. 19-27.

[33] Desiraju, Gautam R.; Vittal, Jagadese J.; Ramanan A. Crystal Engineering: A Textbook. Singapore: World Scientific Publishing Co. Pte. Ltd.; 2011. p. 99-130.

[34] Williams DE. PCK83, QCPE Program 548; Quantum Chemistry Program Exchange. Bloomington, IN: Chemistry department, Indiana University; 1983.

[35] Groom CR, Bruno IJ, Lightfoot MP, Ward SC. The Cambridge structural database. Acta Crystallogr Sect B Struct Sci Cryst Eng Mater. 2016;72:171-9.

[36] Lommerse JPM, Motherwell WDS, Ammon HL, Dunitz JD, Gavezzotti A, Hofmann DWM, et al. A test of crystal structure prediction of small organic molecules. Acta Crystallogr Sect B Struct Sci. 2000;56:697-714.

[37] Graser J, Kauwe SK, Sparks TD. Machine Learning and Energy Minimization Approaches for Crystal Structure Predictions: A Review and New Horizons. Chem Mater. 2018;30:3601-12.

[38] Sharma CVK. Crystal Engineering - Where Do We Go from Here? Cryst Growth Des. 2002;2:465-74.

[39] Carlucci L, Ciani G, Proserpio DM, Sironi A. Interpenetrating Diamondoid Frameworks. J Chem Soc, Chem Commun. 1994;2755-6.

[40] Krishnamohan Sharma C V., Broker GA, Rogers RD. Polymorphous one-dimensional tetrapyridylporphyrin coordination polymers which

structurally mimic aryl stacking interactions. J Solid State Chem. 2000;152:253-60.

[41] Hennigar TL., MacQuarrie DC., Losier P, Rogers RD., Zaworotko MJ. Supramolecular Isomerism in Coordination Polymers: Conformational Freedom. Angew Chemie Int Ed English. 1997;36:972-3.

[42] Moulton B, Zaworotko MJ. From molecules to crystal engineering: Supramolecular isomerism and polymorphism in network solids. Chem Rev. 2001;101:1629-58.

[43] Polymorphism S. Guidance for industry. ANDAs: Pharmaceutical Solid Polymorphism. Chemistry, Manufacturing and Controls Information [Internet]. 2007. Available from: http://www.fda.gov/downloads/drugs/guidancecomplianceregulatory information/guidances/ucm072866.pdf

[44] Aitipamula S, Banerjee R, Bansal AK, Biradha K, Cheney ML, Choudhury AR, et al. Polymorphs, salts, and cocrystals: What's in a name? Cryst Growth Des. 2012;12:2147-52.

[45] Nangia A, Desiraju GR. Pseudopolymorphism: Occurrences of hydrogen bonding organic solvents in molecular crystals. Chem Commun. 1999;605-6.

[46] Seddon KR. Pseudopolymorph: A polemic. Cryst Growth Des. 2004;4:1087.

[47] Nangia A. Pseudopolymorph: Retain this widely accepted term. Cryst Growth Des. 2006;6:2-4.

[48] Robin AY, Fromm KM. Coordination polymer networks with O- and N-donors: What they are, why and how they are made. Coord Chem Rev. 2006;250:2127-57.

[49] J. C. Bailar J. "Coordination Polymers," A Chapter in "Inorganic Polymers, An International Symposium,." In London: The Chemical Society; 1961. p. 51-66.

[50] Janiak C, Hemling H. Linkage lsomerism in bis[hydrotris(1,2,4-triazolyl)borato]zinc from a molecular chelate complex to a three-dimensional co-ordination polymer. J Chem Soc, Dalt Trans. 1994;2947-52.

[51] Biradha K, Zaworotko MJ. supramolecular isomerism and polymorphism in dianion salts of pyromellitic acid. Cryst Eng. 1998;1:67-78.

[52] Soldatov D V., Ripmeester JA, Shergina SI, Sokolov IE, Zanina AS, Gromilov SA, et al. α- And β-bis(1,1,1-trifluoro-5,5-dimethyl-5-methoxyacetylacetonato)copper(II): Transforming the dense polymorph into a versatile new microporous framework. J Am Chem Soc. 1999;121:4179-88.

[53] Swift JA, Pivovar AM, Reynolds AM, Ward MD. Template-directed architectural isomerism of open molecular frameworks: Engineering of crystalline clathrates. J Am Chem Soc. 1998;120:5887-94.

[54] Makal TA, Yakovenko AA, Zhou HC. Isomerism in metal-organic frameworks: "Framework isomers." J Phys Chem Lett. 2011;2:1682-9.

[55] Buerger MJ. Phase Transformations in Solids. Smoluchowski R, Mayer JE, Weyl WA, editors. New York, USA: Wiley; 1951. p. 183-211.

[56] Burger A, Ramberfer R. On the Polymorphism of Pharmaceuticals and Other Molecular Crystals. I. Mikrochim Acta. 1979;72:259-71.

[57] Burger A, Ramberger R. On the polymorphism of pharmaceuticals and

other molecular crystals. II. Mikrochim Acta. 1979;72:273-316.

[58] Brittain HG. Computational methodologies: Toward crystal structure and polymorph prediction. In: Polymorphism in Pharmaceutical Solids. Second Edi. New York, USA: CRC Press; 2016. p. 52-75.

[59] Mullin JW. Crystallization. Fourth Edi. Oxford, UK: Butterworth Heinemann; 2001. p. 181-189.

[60] Dunitz JD, Bernstein J. Disappearing Polymorphs. Acc Chem Res. 1995;28:193-200.

[61] Richardson MF, Yang Q -C, Novotny-Bregger E, Dunitz JD. Conformational polymorphism of dimethyl 3,6-dichloro-2,5-dihydroxyterephthalate. II. Structural, thermodynamic, kinetic and mechanistic aspects of phase transformations among the three crystal forms. Acta Crystallogr Sect B. 1990;46:653-60.

[62] Iasco O, Boillot ML, Bellec A, Guillot R, Rivière E, Mazerat S, et al. The disentangling of hysteretic spin transition, polymorphism and metastability in bistable thin films formed by sublimation of bis(scorpionate) Fe(II) molecules. J Mater Chem C. 2017;5:11067-75.

[63] Goodwin CAP, Ortu F, Reta D, Chilton NF, Mills DP. Molecular magnetic hysteresis at 60 kelvin in dysprosocenium. Nature. 2017;548:439-42.

[64] Eggeler G, Hornbogen E, Yawny A, Heckmann A, Wagner M. Structural and functional fatigue of NiTi shape memory alloys. Mater Sci Eng A. 2004;378:24-33.

[65] Norfleet DM, Sarosi PM, Manchiraju S, Wagner MFX, Uchic MD, Anderson PM, et al. Transformation-induced plasticity during pseudoelastic deformation in Ni-Ti microcrystals. Acta Mater. 2009;57:3549-61.

[66] Straka L, Drahokoupil J, Veřtát P, Kopeček J, Zelený M, Seiner H, et al. Orthorhombic intermediate phase originating from {110} nanotwinning in Ni50.0Mn28.7Ga21.3 modulated martensite. Acta Mater. 2017;132:335-44.

[67] Zhang Z, James RD, Müller S. Energy barriers and hysteresis in martensitic phase transformations. Acta Mater. 2009;57:4332-52.

[68] Flanagan TB., Park C-N., Oates WA. Hysteresis in solid-state reactions. Prog Solid State Chem. 1995;23:291-363.

[69] Patterson JP, Xu Y, Moradi MA, Sommerdijk NAJM, Friedrich H. CryoTEM as an Advanced Analytical Tool for Materials Chemists. Acc Chem Res. 2017;50:1495-501.

[70] Liu L, Chen Z, Wang J, Zhang D, Zhu Y, Ling S, et al. Imaging defects and their evolution in a metal–organic framework at sub-unit-cell resolution. Nat Chem. 2019;11:622-8.

[71] Lyu J, Gong X, Lee SJ, Gnanasekaran K, Zhang X, Wasson MC, et al. Phase Transitions in Metal-Organic Frameworks Directly Monitored through in Situ Variable Temperature Liquid-Cell Transmission Electron Microscopy and in Situ X-ray Diffraction. J Am Chem Soc. 2020;142:4609-15.

[72] Parent LR, Pham CH, Patterson JP, Denny MS, Cohen SM, Gianneschi NC, et al. Pore Breathing of Metal–Organic Frameworks by Environmental Transmission Electron Microscopy. J Am Chem Soc. 2017;139:13973-6.

[73] Kolb U, Mugnaioli E, Gorelik TE. Automated electron diffraction tomography - A new tool for

nano crystal structure analysis. Cryst Res Technol. 2011;46:542-54.

[74] Wan W, Sun J, Su J, Hovmöller S, Zou X. Three-dimensional rotation electron diffraction: Software RED for automated data collection and data processing. J Appl Crystallogr. 2013;46:1863-73.

[75] Wang B, Rhauderwiek T, Inge AK, Xu H, Yang T, Huang Z, et al. A Porous Cobalt Tetraphosphonate Metal–Organic Framework: Accurate Structure and Guest Molecule Location Determined by Continuous-Rotation Electron Diffraction. Chem - A Eur J. 2018;24:17429-33.

[76] Hutchison IB, Bull CL, Marshall WG, Urquhart AJ, Oswald IDH. Pressure-induced polymorphism of caprolactam: A neutron diffraction study. Molecules. 2019;24:2174.

[77] Smokrović K, Muratović S, Karadeniz B, Užarević K, Žilić D, Dilović I. Synthon Robustness and Structural Modularity of Copper(II) Two-Dimensional Coordination Polymers with Isomeric Amino Acids and 4,4′-Bipyridine. Cryst Growth Des. 2020;20:2415-23.

[78] Fan WW, Cheng Y, Zheng LY, Cao QE. Reversible Phase Transition of Porous Coordination Polymers. Chem - A Eur J. 2020;26:2766-79.

[79] Hoffman AEJ, Vanduyfhuys L, Nevjestić I, Wieme J, Rogge SMJ, Depauw H, et al. Elucidating the Vibrational Fingerprint of the Flexible Metal-Organic Framework MIL-53(Al) Using a Combined Experimental/ Computational Approach. J Phys Chem C. 2018;122:2734-46.

[80] Dudek MK, Kazmierski S, Kostrzewa M, Potrzebowski MJ. Solid-State NMR Studies of Molecular Crystals. In: Annual Reports on NMR Spectroscopy. Elsevier Ltd; 2018. p. 1-81.

[81] Shin CH, Huh JO, Lee MH, Do Y. Polymorphism-induced dual phosphorescent emission from solid-state iridium(III) complex. J Chem Soc Dalt Trans. 2009;6476-9.

[82] Quartapelle Procopio E, Mauro M, Panigati M, Donghi D, Mercandelli P, Sironi A, et al. Highly emitting concomitant polymorphic crystals of a dinuclear rhenium complex. J Am Chem Soc. 2010;132:14397-9.

[83] Schaefer J, Stejskal EO. Carbon-13 Nuclear Magnetic Resonance of Polymers Spinning at the Magic Angle. J Am Chem Soc. 1976;98:1031-2.

[84] Zhou J, Kye YS, Harbison GS. Isotopomeric polymorphism. J Am Chem Soc. 2004;126:8392-3.

[85] Karadeniz B, Žilić D, Huskić I, Germann LS, Fidelli AM, Muratović S, et al. Controlling the Polymorphism and Topology Transformation in Porphyrinic Zirconium Metal-Organic Frameworks via Mechanochemistry. J Am Chem Soc. 2019;141:19214-20.

[86] Nehrkorn J, Holldack K, Bittl R, Schnegg A. Recent progress in synchrotron-based frequency-domain Fourier-transform THz-EPR. J Magn Reson. 2017;280:10-9.

[87] Novikov V V., Pavlov AA, Nehrkorn J, Nelyubina Y V. Influence of Polymorphism on the Magnetic Properties of Single-Molecule Magnets According to the Data of EPR Spectroscopy in the Terahertz Range. Russ J Coord Chem Khimiya. 2020;46:756-61.

[88] Voloshin Y, Belaya I, Krämer R. Cage Metal Complexes. Clathrochelates Revisited. Cham, Switzerland: Springer Nature; 2017. p. 231-317.

[89] Schneemann A, Bon V, Schwedler I, Senkovska I, Kaskel S, Fischer RA. Flexible metal-organic

frameworks. Chem Soc Rev. 2014;43:6062-96.

[90] Longley L, Li N, Wei F, Bennett TD. Uncovering a reconstructive solid–solid phase transition in a metal–organic framework. R Soc Open Sci. 2017;4.

[91] Thompson JR, Katz MJ, Williams VE, Leznoff DB. Structural Design Parameters for Highly Birefringent Coordination Polymers. Inorg Chem. 2015;54:6462-71.

[92] Sato K, Boistelle R. Stability and occurrence of polymorphic modifications of the stearic acid in polar and nonpolar solutions. J Cryst Growth. 1984;66:441-50.

[93] Hulliger J. Chemistry and Crystal Growth. Angew Chemie Int Ed English. 1994;33:143-62.

[94] Carson F, Su J, Platero-Prats AE, Wan W, Yun Y, Samain L, et al. Framework isomerism in vanadium metal-organic frameworks: MIL-88B(V) and MIL-101(V). Cryst Growth Des. 2013;13:5036-44.

[95] Tanasaro T, Adpakpang K, Ittisanronnachai S, Faungnawakij K, Butburee T, Wannapaiboon S, et al. Control of Polymorphism of Metal-Organic Frameworks Using Mixed-Metal Approach. Cryst Growth Des. 2018;18:16-21.

[96] Kukovec BM, Venter GA, Oliver CL. Structural and DFT studies on the polymorphism of a cadmium(II) dipicolinate coordination polymer. Cryst Growth Des. 2012;12:456-65.

[97] Hall SR, Hall CL, Guo R, Potticary J, Cremeens ME, Warren SD, et al. Color differences highlight concomitant polymorphism of chalcones. Cryst Growth Des. 2020;20:6346-55.

[98] Langenstroer A., Kartha KK., Dorca Y, Droste J, Stepanenko V, Albuquerque RQ., et al. Unraveling Concomitant Packing Polymorphism in Metallosupramolecular Polymers. J Am Chem Soc. 2020;141:5192-200.

[99] Speight IR, Huskić I, Arhangelskis M, Titi HM, Stein RS, Hanusa TP, et al. Disappearing Polymorphs in Metal–Organic Framework Chemistry: Unexpected Stabilization of a Layered Polymorph over an Interpenetrated Three-Dimensional Structure in Mercury Imidazolate. Chem - A Eur J. 2020;26: 1811-8.

[100] Masciocchi N, Ardizzoia GA, Brenna S, Castelli F, Galli S, Maspero A, et al. Synthesis and ab-initio XRPD structure of group 12 imidazolato polymers. Chem Commun. 2003;3: 2018-9.

[101] Bond AD, Boese R, Desiraju GR. On the polymorphism of aspirin: Crystalline aspirin as intergrowths of two "polymorphic" domains. Angew Chemie - Int Ed. 2007;46:618-22.

[102] Bürgi HB, Hostettler M, Birkedal H, Schwarzenbach D. Stacking disorder: The hexagonal polymorph of tris(bicyclo[2.1.1]hexeno) benzene and related examples. Zeitschrift fur Krist. 2005;220:1066-75.

[103] Mishra MK, Desiraju GR, Ramamurty U, Bond AD. Studying microstructure in molecular crystals with nanoindentation: Intergrowth polymorphism in felodipine. Angew Chemie - Int Ed. 2014;53:13102-5.

[104] Carlucci L, Ciani G, Proserpio DM. A new type of entanglement involving one-dimensional ribbons of rings catenated to a three-dimensional network in the nanoporous structure of $[Co(bix)_2(H_2O)_2](SO_4){\cdot}7H_2O$ [bix = 1,4-bis(imidazol-1-ylmet. Chem Commun. 2004;4:380-1.

[105] Carlucci L, Ciani G, Proserpio DM, Spadacini L. Supramolecular isomers in the same crystal: a new case

involving two different types of layers polycatenated in the 3D architecture of [Cu(bix)$_2$(SO$_4$)]·7.5H$_2$O[bix=1,4-bis(imidazol-1-ylmethyl)benzene]. CrystEngComm. 2004;6:96-101.

[106] Liu G, Gou R, Li H, Zhang C. Polymorphism of Energetic Materials: A Comprehensive Study of Molecular Conformers, Crystal Packing, and the Dominance of Their Energetics in Governing the Most Stable Polymorph. Cryst Growth Des. 2018;18:4174-86.

[107] Braga D, Brammer L, Champness NR. New trends in crystal engineering. CrystEngComm. 2005;7:1-19.

[108] Bernstein J. Polymorphism and Patents from a Chemist's Point of View. In: Polymorphism: in the Pharmaceutical Industry. Hilfiker R, editor. Weinheim, Germany: Wiley-VCH; 2006. p. 369-72.

[109] Roelands CPM, Jiang S, Kitamura M, ter Horst JH., Kramer HJM., Jansens PJ. Antisolvent Crystallization of the Polymorphs of L-Histidine as a Function of Supersaturation Ratio and of Solvent Composition. 2006;6:955-63.

[110] Sunday NF. Emerging Trends in Coordination Polymers and Metal-Organic Frameworks: Perspectives, Synthesis, Properties and Applications. Arch Org Inorg Chem Sci. 2018;1:39-51.

[111] Czaja AU, Trukhan N, Müller U. Industrial applications of metal–organic frameworks. Chem Soc Rev. 2009;38:1284-93.

[112] Hartmann M, Böhme U, Hovestadt M, Paula C. Adsorptive Separation of Olefin/Paraffin Mixtures with ZIF-4. Langmuir. 2015;31:12382-9.

[113] Bennett TD, Saines PJ, Keen DA, Tan JC, Cheetham AK. Ball-milling-induced amorphization of zeolitic imidazolate frameworks (ZIFs) for the irreversible trapping of iodine. Chem - A Eur J. 2013;19:7049-55.

[114] Kitagawa S, Kitaura R, Noro SI. Functional porous coordination polymers. Angew Chemie - Int Ed. 2004;43:2334-75.

[115] Biradha K, Goswami A, Moi R. Coordination polymers as heterogeneous catalysts in hydrogen evolution and oxygen evolution reactions. Chem Commun. 2020;56:10824-42.

[116] Li H, Eddaoudi M, O'Keeffe M., Yaghi OM. Design and synthesis of an exceptionally stable and highly. Nature. 1999;402:276-9.

[117] Widmer RN, Lampronti GI, Chibani S, Wilson CW, Anzellini S, Farsang S, et al. Rich Polymorphism of a Metal-Organic Framework in Pressure-Temperature Space. J Am Chem Soc. 2019;141:9330-7.

[118] Tu M, Kravchenko DE, Xia B, Rubio-Giménez V, Wauteraerts N, Verbeke R, et al. Template-mediated control over polymorphism in the vapor-assisted formation of zeolitic imidazolate framework powders and films. Angew Chemie Int Ed. 2021;60:1-7.

[119] Erk, Peter; Hengelsberg H. Applications of Phthalocyanines. In: The Porphyrin Handbook. Kadish KM, Smith KM, Guilard R, editors. San Diego, California, USA: Academic Press; 2003. p. 118-22.

[120] Caradoc-Daviesa PL, Hanto LR, Henderson W. Coordination polymers and isomerism; a study using silver(I) and a π-stacked ligand. J Chem Soc Dalt Trans. 2001;2749-55.

Chapter 7

Brushite: Synthesis, Properties, and Biomedical Applications

Khalil Issa, Abdulaziz Alanazi, Khalid A. Aldhafeeri, Ola Alamer and Mazen Alshaaer

Abstract

In this chapter, besides its biomedical applications, the synthesis and properties of brushite were investigated. Brushite consists of two types of crystals, platy and needle-like, and their formation depends on the pH of the medium during precipitation. Platy crystals are formed in a slightly acidic medium, pH = 5, and needle-like crystals at a higher pH = 6.5–7. In this study, the monoclinic brushite crystals were synthesized using dissolution-precipitation reactions. It is found that the brushite crystal growth occurs mainly along the (020) crystallographic plane. The thermogravimetric analysis confirms the presence of the two structural water molecules, which decompose at a temperature range between 80 and 220°C. Brushite was used in the preparation of tetracalcium phosphate mineral, which is the powder component for calcium phosphate cement (CPC). CPC was subsequently prepared from TTCP and phosphate-based hardening solution. *In vitro* evaluation of the resultant CPC using Hanks' Balanced Salt Solution results in the growth of nanofibrous crystals of Calcium-deficient hydroxyapatite (CDHA) layers on the surfaces of the CPC. The cultured CPC exhibits new connective tissues and throughout the CaP matrix.

Keywords: brushite, hydroxyapatite, tetracalcium phosphates, bioactivity, porosity

1. Introduction

Calcium phosphates (CaP) are one of the most important compounds found in nature [1, 2]. There are many applications for these compounds in various fields, especially agricultural, environmental, and medical applications. CaP are characterized by their wide diversity as it is produced at different temperatures and pH ranges. In addition, the molar ratio of calcium to phosphorous in the precursors plays an important role in the resulting materials. The biochemical characteristics and mineralogical structures of CaP are similar to inorganic constituents of mammals' bones [3, 4]. Because of their excellent biocompatibility, high bioactivity, and low toxicity, CaP are considered as a good candidate for bone tissue engineering applications. The CaP minerals such as hydroxyapatite, brushite, tricalcium phosphates (TCP) are used as precursors for the preparation of bone cements and bio-ceramics [5, 6]. Therefore, these minerals are widely used for biomedical applications such as drug delivery and bone tissue engineering (**Table 1**) [6–9].

CaP belong to the family of apatite. There are several CaP phases, the most ubiquitous being hydroxyapatite [HAp, $Ca_{10}(PO_4)_6(OH)_2$]. Other CaP structures include brushite (DCPD, $CaHPO_4{\cdot}2H_2O$) and tricalcium phosphate

Ca/P ratio	Compound	Formula	Density (g/cm^3)	pH stability range (25°C) [a]
0.5	monocalcium phosphate monohydrate (MCPM)	$Ca(H_2PO_4)_2{\cdot}H_2O$	2.23	0.0–2.0
0.5	monocalcium phosphate anhydrate (MCPA)	$Ca(H_2PO_4)_2$	2.58	[d]
1.0	dicalcium phosphate dihydrate (DCPD, brushite)	$CaHPO_4{\cdot}H_2O$	2.32	2.0–6.0
1.0	dicalcium phosphate anhydrate (DCPA, monetite)	$CaHPO_4$	2.89	[d]
1.0	Calcium pyrophosphate (pyro)	$Ca_2P_2O_7$	–	–
1.33	octacalcium phosphate (OCP)	$Ca_8(HPO_4)_2(PO_4)_4{\cdot}5H_2O$	2.61	5.5–7.0
1.5	α-tricalcium phosphate (α-TCP)	α -$Ca_3(PO_4)_2$	2.86	[b]
1.5	β-tricalcium phosphate (β-TCP)	β -$Ca_3(PO_4)_2$	3.07	[b]
1.2–2.2	amorphous calcium phosphate (ACP)	$Ca_x(PO_4)y{\cdot}nH_2O$	—	[c]
1.5–1.67	calcium-deficient hydroxyapatite (CDHA)	$Ca_{10-x}(HPO_4)_x(PO_4)_{6-x}(OH)_{2-x}$ (0 < x < 1)		6.5–9.5
1.67	hydroxyapatite (HAp)	$Ca_{10}(PO_4)_6(OH)_2$	3.15	9.5–12
2.0	tetracalcium phosphate (TTCP)	$Ca_4(PO_4)_2O$	3.05	[b]

[a] In aqueous solution, [b] These compounds cannot be precipitated from aqueous solutions. [c] Cannot be measured precisely. However, the following values were reported: 25.7 ± 0.1 (pH 7.40), 29.9 ± 0.1 (pH 6.00), and 32.7 ± 0.1 (pH 5.28). [d] Stable at temperatures above 100°C.

Table 1.
List of main phases of CaP [6–9].

(TCP, $Ca_3(PO_4)_2$). Several low- and high-temperature approaches have been reported for synthesizing HAp and brushite (DCPD), while TCP is primarily synthesized using high-temperature methods [6]. The chemical formation of CaP minerals is common in natural systems, although the elucidation of the mechanisms of formation and transformations between the crystal forms of the minerals remains a major challenge. The most thermodynamically stable form, at ambient temperature and pressure, is calcium hydroxyapatite (HAp); however, this does not form readily without a transition phase. Other mineral phases, such as octacalcium phosphate (OCP) and amorphous TCP, are precursor phases that can transform to HAp [10].

Acidic CaP, such as brushite (dicalcium phosphate dehydrate, DCPD), are thermodynamically unstable under pH values greater than 6–7 and thus undergo transformation into more stable CaP. Researchers have also demonstrated that meta-stable brushite (DCPD) may convert to OCP or calcium hydroxyapatite (HAp), and that OCP may convert to hydroxyapatite, depending on the Ca/P ratio and the pH value of the setting reactions [11]. Brushite, a type of CaP that is the most easily synthesized, transforms into monetite (dicalcium phosphate anhydrate, DCPA) at temperatures above 80°C. Monetite (DCPA) is the anhydrous form of brushite (DCPD) and can, like brushite, be crystallized from aqueous solutions, but only when the temperature is above 80°C. At low pH values (<7), monetite is the most stable of the CaP, although the conversion of brushite to monetite is faster when the water is warmer and more acidic [8].

Brushite-based biomaterials are characterized by good bioactivity, and they are bioresorbable and biocompatible. Unlike apatite-based materials, brushite-based ones are rapidly resorbed in vivo [12]. The bioactivity and biocompatibility of brushite-based biomaterials have been investigated in several compositions, applications, and in vivo [11]. Brushite-based materials are biocompatible with and tolerated by soft tissues and bone in vivo, so that material resorption was shortly followed by the formation of new bone tissues. Histological measurements and experimental studies indicate that brushite-based materials feature good biocompatibility, with no appearance of inflammatory cells [12].

Amorphous calcium phosphate (ACP) is often encountered as a transient phase during the formation of CaP in aqueous systems. Usually, ACP is the first phase that is precipitated from a supersaturated solution prepared by the rapid mixing of solutions containing calcium cations and phosphate anions. The chemical composition of ACP strongly depends on the solution's pH value and the concentrations of calcium and phosphate ions. For example, ACP phases are formed with Ca/P ratios in the range of 1.18:1 (precipitated in a solution with a pH value of 6.6) to 1.53:1 (precipitated in a solution with a pH value of 11.7), although ratios of up to 2.5:1 have also been encountered [8]. The structure of ACP is still uncertain and it has been reported to be more soluble than brushite [13]. ACP could be stabilized by another chemical compound: e.g., pyrophosphate ($P_2O_7^{4-}$) retards the conversion of ACP to apatite. Finally, ACP is characterized by its relatively high solubility and ability to obtain a substantial release of Ca^{2+} and PO_4 ions [12].

This chapter aims to synthesis brushite as one of the most common CaP. The microstructural and thermal properties of this mineral are characterized and discussed. After that, the thermochemical transformation of brushite was investigated. The resultant compound is used as precursors for bone cement. Finally, biomineralization and the bioactivity, and biomineralization of calcium phosphate cement (CPC) are studied *in vitro*.

2. Synthesis of brushite ($CaHPO_4{\cdot}2H_2O$)

The typical method synthesis of the brushite powder was performed at ambient according to the following equation:

$$Ca(NO_3)_2 + Na_2HPO_4 \rightarrow \mathrm{CaHPO_4{\cdot}2H_2O + 2NaNO_3} \quad (1)$$

Two solutions were prepared for the synthesis of brushite. The first solution is prepared by dissolving 0.5 mol of $Na_2HPO_4{\cdot}2H_2O$ in 1 L of distilled water, and the second solution is a result of dissolving 0.5 mol of $Ca(NO_3).6H_2O$ in 1 L of distilled water. After preparing the solutions, 200 ml of the $Ca(NO_3)_2$ was added dropwise using a glass funnel with a glass stopcock (flow rate is 2 ml/min) to the $Na_2HPO_4{\cdot}2H_2O$ solution while stirring and adjusting the pH between 6 and 6.5 using ammonia solution (25%, Labochemie, India). Afterward, the resultant solution with precipitates was stirred (400 rpm) at ambient conditions for 1 hour to ensure a homogeneous mixture. The precipitate was vacuum filtered using a qualitative filter paper via a Buchner funnel, washed three times with de-ionized water and another three times with ethanol to reduce the possibility of agglomeration [14, 15], after which it was placed upon a watch glass and dried at 40°C overnight in a drying-oven. After the formation of the precipitate, some of the powder was washed with distilled water, then dried using ethanol at 40°C for a week. The major steps of the experimental design are reported in **Figure 1**.

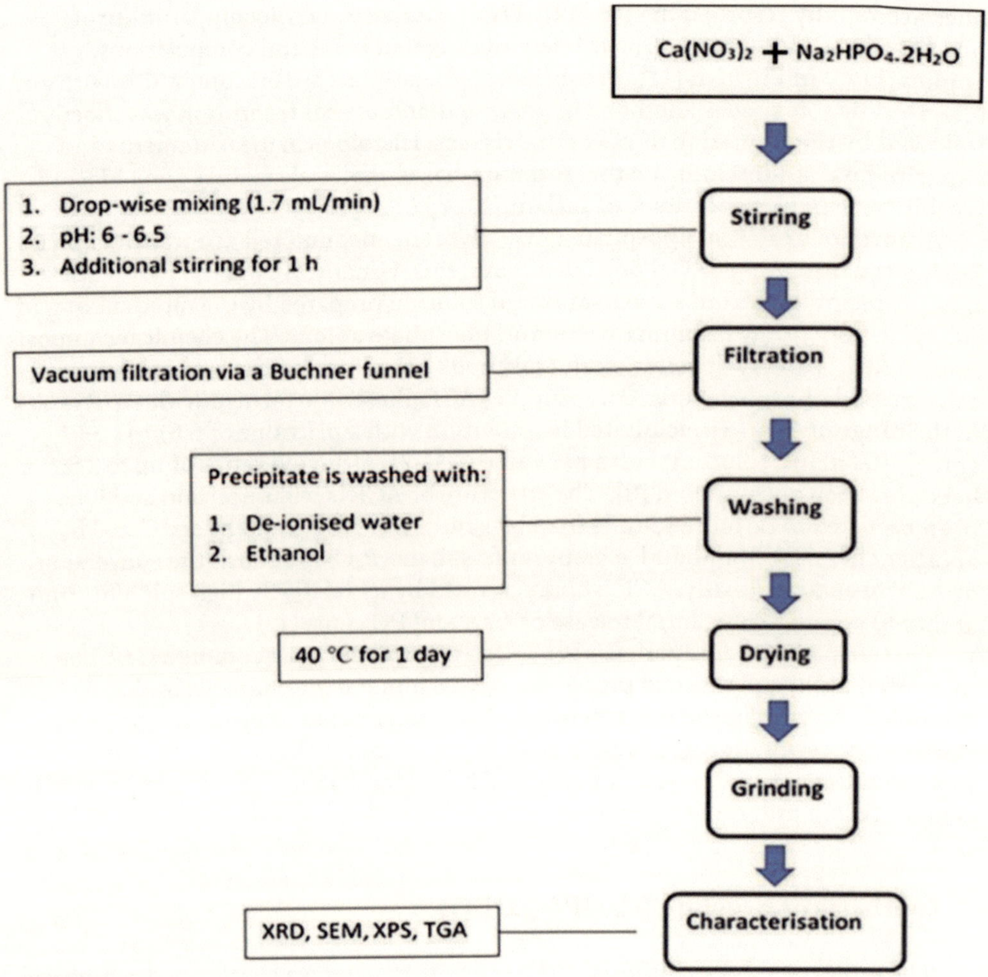

Figure 1.
preparation of brushite [16].

3. Crystal morphology of brushite

Two morphologies of brushite crystal layers are reported in our previous work [17]. These brushite crystals consist of [2] platy particles (**Figure 2A**) and a continuous needle-like (**Figure 2B**) dendritic network [18]. The plate-like brushite crystals mostly formed in parallelogram shapes stacked in multiple layers. Their dimensions ranged from 1 μm to a few micrometers, in two directions. The thickness of the platy crystals is measured in the nano-scale. The needle-like brushite particles appeared to have a dendritic network structure. The total length of the crystal is around 40 μm. The EDS measurements (**Figure 2C**) showed that the needle-like brushite crystals are composed of, by atomic percentage: O (76.46%), Ca (12.04%), and P (10.95%). The fact that the percentages of P and Ca are nearly equal and is in good agreement with the theoretical Ca/P atomic percentages of brushite [19, 20]. The brushite morphology depends on the pH of the solution during the precipitation; at acidic pH, around 5, platy crystals of brushite are formed (**Figure 2A**), whereas needle-like (**Figure 2B**) ones are predominant at a higher pH [2].

SEM images of the brushite crystals are shown in **Figure 3**. These plate-like crystals are obtained according to the experimental procedure as reported in **Figure 1**.

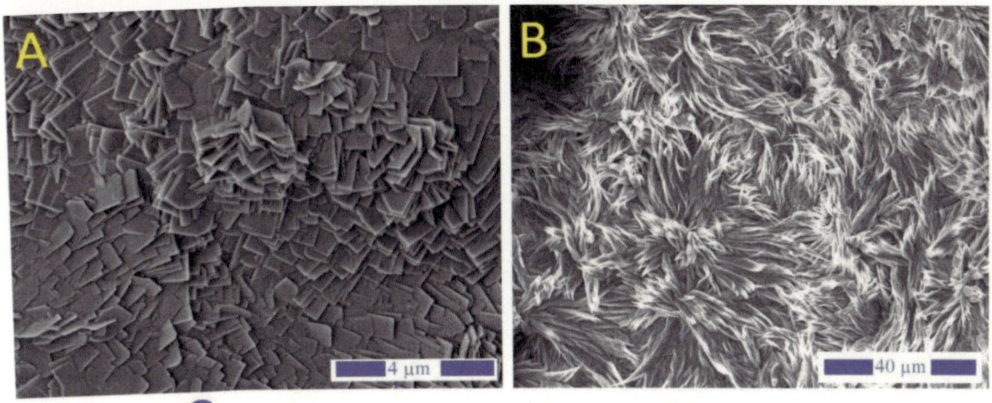

C

El	AN	Series	unn. C [wt.%]	norm. C [wt.%]	Atom. C [at.%]	Error [%]
O	8	K-series	58.66	59.45	76.46	7.0
Na	11	K-series	0.61	0.62	0.55	0.1
P	15	K-series	16.26	16.48	10.95	0.6
Ca	20	K-series	23.15	23.46	12.04	0.8

Figure 2.
(A) crystal morphologies of brushite; platy structure, (B) needle-like structure, and (C) EDS analysis [17].

Figure 3.
*SEM image of monoclinic brushite crystals, see **Figure 1** (preparation procedure).*

It is known that the morphology of brushite is characterized by a plate-like or needle-like structure, depending on the solution pH used [11, 13]. The plate-like crystals are thin (~400 nm), while their width and elongation are approximately 10 and 20 μm, respectively, values similar to those reported in other studies [21].

The XRD patterns of brushite, as well as the patterns of standard brushite, are shown in **Figure 4**. The mineralogy of the powder confirms that this precipitate produced after mixing $NaH_2PO_4{\cdot}2H_2O$ and $Ca(NO_3)_2{\cdot}4H_2O$ solutions with a Ca:P molar ratio 1:1 (**Figure 1**) is pure brushite, while its crystals grow after nucleation in proportion to the three major planes, namely, (020), (121-), and (141). All peaks of the powder pattern denote the brushite's monoclinic structure [16], while the peak at 11.7° 2-Theta indicates that the crystal growth takes place primarily along the (020) crystallographic plane. Rietveld refined unit cell parameters for brushite are presented in **Table 2**.

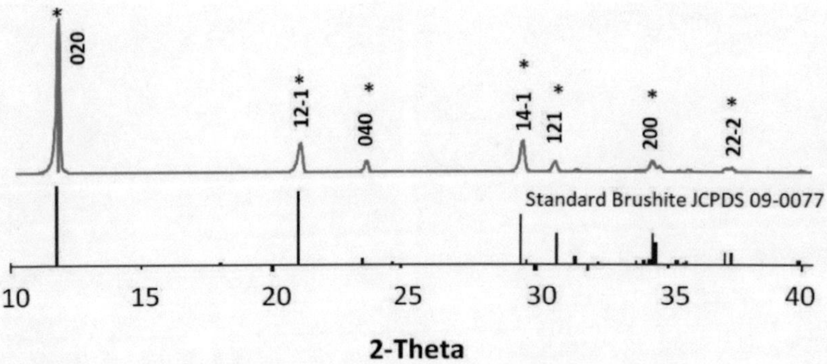

Figure 4.
XRD patterns of brushite (synthesis details are shown in ***Figure 1****) [16].*

wt%	a(Å)	b(Å)	c(Å)	(βo)	V(Å3)
100.0	5.8145	15.1693	6.2399	116.392	492.83

*Standard deviation varied between 0.001 and 0.005 for all samples.

Table 2.
*Refined unit cell parameters for brushite from XRD data using the Rietveld approach.**

4. Thermal properties of brushite

The results of the TG analysis for the brushite are reported in **Figure 5**. Brushite is considered as a water-bearing phosphate mineral [2] and its crystal structure contains compact sheets consisting of parallel chains in which Ca ions are coordinated by six phosphate ions and two oxygen atoms belonging to the water molecules [22].

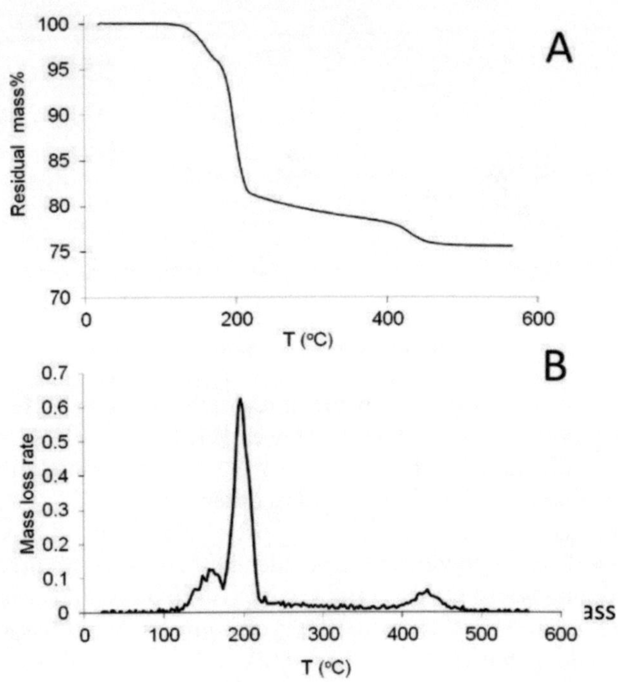

Figure 5.
TGA of brushite: (A) cumulative mass loss, and (B) mass loss rate.

Brushite contains two water molecules in its lattice and adsorbed water molecules on its surface, as indicated by the presence of two sharp peaks of mass loss during heating between 80 and 220°C (**Figure 5A**) [2, 23]. Part of the chemically-bound water is released during the transformation of brushite to monetite, $CaHPO_4$, at ~220°C [17], and later to calcium pyrophosphate, $Ca_2P_2O_7$, at ~400°C [24]. Pyrophosphates are decomposed at higher temperatures of 750–800°C (**Figure 5B**) [8, 25]. The heating of pure brushite to 600°C results in a mass loss of approximately 25%wt, while the theoretical mass loss for the dehydration of brushite is 20.93%wt [1].

Dehydration of brushite over the temperature range 110–215°C takes place according to Eq. (2) and normally results in a weight loss of about 19%wt, while the formation of calcium pyrophosphate is accomplished by Eq. (3) a

$$CaHPO_4{\cdot}2H_2O \rightarrow CaHPO_4 + 2H_2O \tag{2}$$

$$2CaHPO_4 \rightarrow Ca_2P_2O_7 + H_2O \tag{3}$$

Figure 5B shows the rate of mass loss as a function of heating temperature for brushite. More specifically, **Figure 5B** shows the dehydration peaks corresponding to the two water molecules of brushite [26, 27].

5. Calcium phosphate cement (CPC)

Brushite is used to prepare the powder component, tetracalcium phosphate (TTCP), of the CPC. Brushite was calcined at 500°C and transformed into a more stable phase; calcium pyrophosphate ($Ca_2P_2O_7$). Afterward, calcium pyrophosphate and calcium carbonate were mixed with a Ca/P molar ratio of 1.9 [5]. These two compounds were mixed in ethanol for 10 h. Then the resultant mixture was dried at 105°C for 24 h and then crushed. The mixture was heated at 1500°C for 4 h quenching to room temperature, see **Figure 7**. The resultant powder (TTCP) was ground into a fine powder [5]. The general equation TTCP synthesis is as follows:

$$CaHPO_4{\cdot}2H_2O + 2CaCO_3\left(1500°C\text{ for }3h\right) \rightarrow Ca_4\left(PO_4\right)_2O + 2CO_2 \tag{4}$$

Mannitol, sizes vary from 100 to 400 μm, was added to the TTCP with the weight ratio of 0.5. Diammonium hydrogen phosphate solution with a concentration of 33.3 wt% was mixed with TTCP-mannitol mixture, with the weight ratio of 0.34 mL (solution)/g (TTCP). After mixing the CPC components for 2 min, the paste was packed in a polycarbonate mold which has an opening of 10 × 10 mm under a pressure of ~1 MPa at 37°C. The hardened samples were then demolded and immersed in Hanks' physiological solution at 37°C for 1 day [28, 29]. The composition of Hanks' physiological solution is reported in **Table 3**.

After mixing the two components of CPC, TTCP, and the hardening solution, TTCP hydrolyses through a dissolution-precipitation reactions resulting in the formation of layers of Ca-deficient hydroxyapatite (CDHA) crystals, which are similar to the mineral component of the bone from (**Figure 6**). This hydrolysis process occurs during the setting reactions, which is confirmed by XRD and SEM (**Figures 7** and **8**). These CDH layers are characterized by wide range distribution of rod-like crystals [30].

The CPC was synthesized for *in vitro* cultivation [5, 31, 32]. Mesenchymal stem cells (MSCs) were seeded on the CPC porous matrix in presence of an osteogenic

Component	Concentration
NaCl (mw: 58.44 g/mol)	0.14 M
KCl (mw: 74.55 g/mol)	0.005 M
$CaCl_2$ (mw: 110.98 g/mol)	0.001 M
$MgSO_4.7H_2O$ (mw: 246.47 g/mol)	0.0004 M
$MgCl_2.6H_2O$ (mw: 203.303 g/mol)	0.0005 M
$Na_2HPO_4.2H_2O$ (mw: 177.99 g/mol)	0.0003 M
KH_2PO_4 (mw: 136.086 g/mol)	0.0004 M
D-Glucose (Dextrose) (mw: 180.156 g/mol)	0.006 M
$NaHCO_3$ (mw: 84.01 g/mol)	0.004 M

Table 3.
Composition of Hanks' physiological solution.

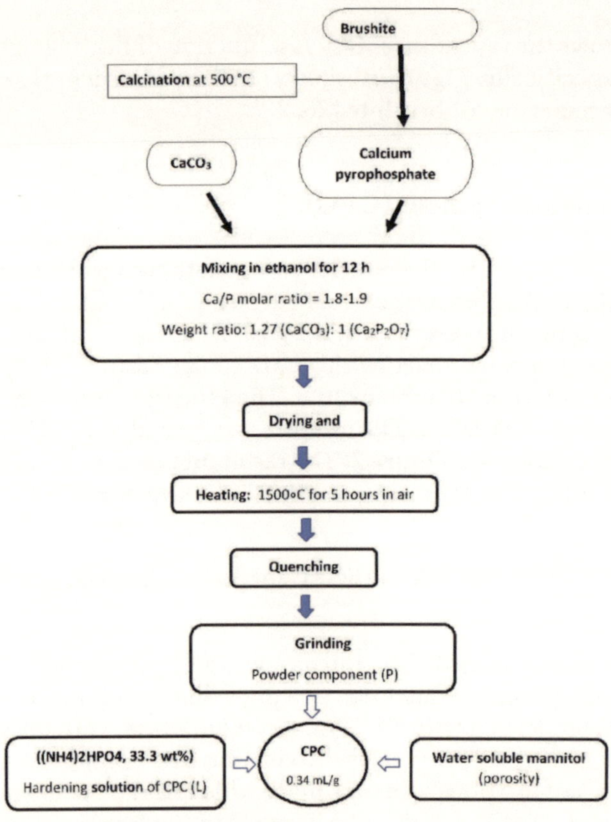

Figure 6.
Schematic diagram of CPC preparation.

medium for 21 days [5, 12]. As a result of *in vitro* cultivation, mineralized nodules were formed in the constructs. The seeded cells grow and their sizes increase from 5 μm to around 50 μm. The growing cells adhered to the CPC matrix and developed cytoplasmic extension as reported in **Figure 8**.

A thick layer of nano fibrous CDH crystals covers the surfaces of the CPC matrix (**Figure 7B**). The cultured CPC exhibits new connective tissues and throughout the CaP matrix (**Figure 8**). The CPC matrix contains bioactive CDH with both Ca and P,

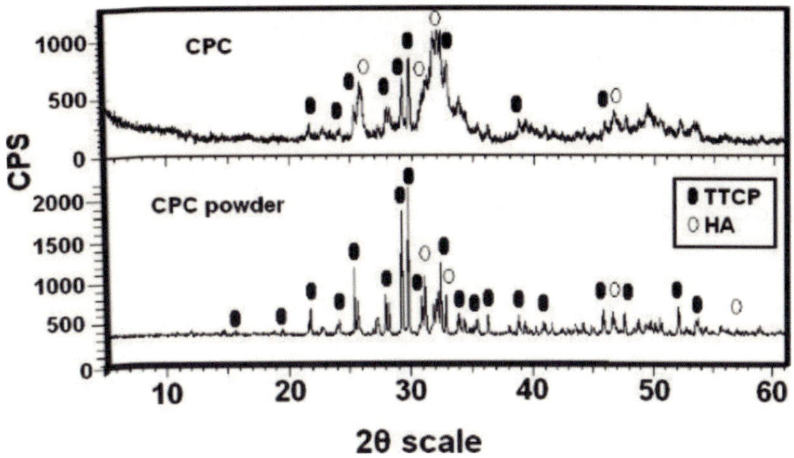

Figure 7.
XRD patterns of CPC and CPC powder [5].

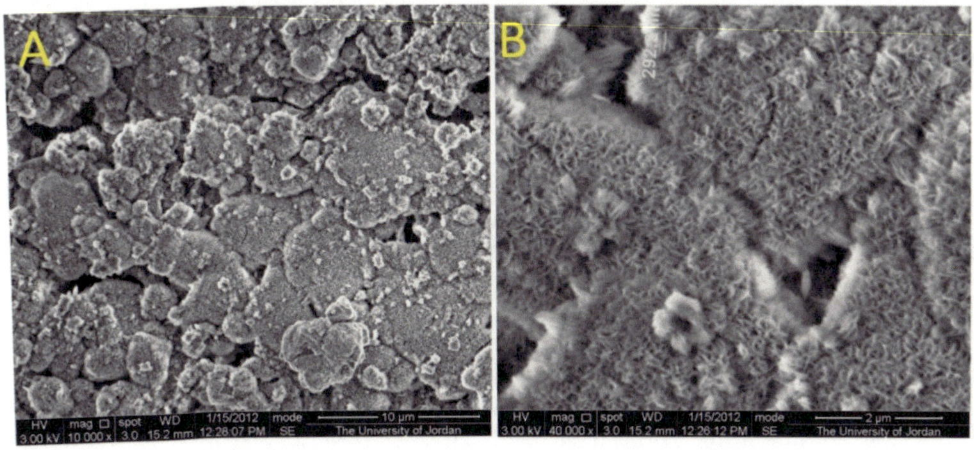

Figure 8.
Growth of thick Ca-deficient hydroxyapatite (CDHA) layer on the surface of the CaP matrix.

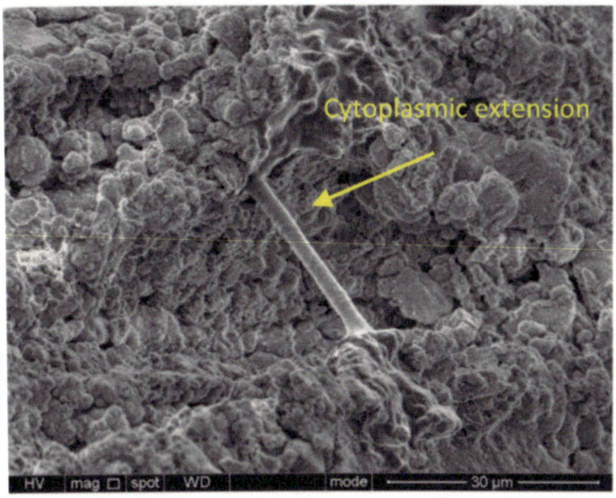

Figure 9.
SEM image of the surface of CPC after MSCs culturing for 21 days.

therefore, this matrix provides the appropriate environment for MSCs growth and osteogenic differentiation (**Figure 9**) [33].

6. Conclusions

Brushite crystals consist of platy and needle-like crystals. It is found that the pH of the solution during the precipitation of brushite plays the main role in determining the shape of the crystals. Usually at a relatively low pH, around 5, platy crystals are formed, while at a higher pH, around ph = 6.5, needle-like crystals are precipitated. In this study, brushite crystals with a monoclinic structure were synthesized using calcium and phosphate salts. The brushite crystal growth occurs mainly along the (020) crystallographic plane. Brushite crystal is characterized by the presence of two structural water molecules. These two molecules are released at a temperature range between 80°C and 220°C to form monetite minerals.

In this chapter, brushite is used as a precursor to synthesize TTCP, the powder components of CPC. As a result of the solid reactions between brushite and calcium carbonate, at high temperature, 1500°C, a new CaP phase is called TTCP. This powder reacts with the phosphate-based solution at 37°C to form CDHA. Immersing this CPC in Hanks' Balanced Salt Solution results in the growth of nanofibrous crystals of CDHA layers on the surfaces of the CPC. The cultured CPC exhibits new connective tissues and throughout the CaP matrix. The CPC matrix contains bioactive CDH with both Ca and P, therefore this matrix provides an appropriate environment for MSCs growth and osteogenic differentiation.

Bioactive features of brushite-based materials affect cell adhesion, proliferation, and new bone formation. Bioactivity can be altered and controlled by the crystal structure and physical property of the scaffold. Bioactive characteristics are different depending on the produced type of CaP phases such as HAP, TCP, and ACP. These different bioactive characteristics are caused by the differences in Ca/P ratio, crystal structure, stability, and solubility. As mentioned above, brushite is often used with other CaP to control and improve their chemical, biological, and physical properties. Various applications have been exploited to actively utilize the bioactive features of brushite in bone regeneration.

Author details

Khalil Issa[1*], Abdulaziz Alanazi[2], Khalid A. Aldhafeeri[2], Ola Alamer[3] and Mazen Alshaaer[2,4]

1 Faculty of Medicine and Health Sciences, Orthopedics Unit, An-Najah National University, Nablus, Palestine

2 Department of Physics, College of Science and Humanities in Al-Kharj, Prince Sattam Bin Abdulaziz University, Al-Kharj, Saudi Arabia

3 Faculty of Applied Science, Department of Physics, Umm Al-Qura University, Makkah, Saudi Arabia

4 GeoBioTec Research Center, University of Aveiro, Campus de Santiago, Aveiro, Portugal

*Address all correspondence to: k.issa@najah.edu

References

[1] Kim Y, Lee SY, Roh Y, Lee J, Kim J, Lee Y, et al. Optimizing calcium phosphates by the control of pH and temperature via wet precipitation. Journal of Nanoscience and Nanotechnology. 2015;**15**(12):10008-10016

[2] Suryawanshi VB, Chaudhari RT. Growth and characterization of agar gel grown brushite crystals. Indian Journal of Materials Science. 2014;**2014**:1-6

[3] Alshaaer M, Kailani MH, Ababneh N, Mallouh SAA, Sweileh B, Awidi A. Fabrication of porous bioceramics for bone tissue applications using luffa cylindrical fibres (LCF) as template. Processing and Application of Ceramics. 2017;**11**(1):13-20

[4] Radwan NH, Nasr M, Ishak RA, Abdeltawa NF, Awad GA. Chitosan-calcium phosphate composite scaffolds for control of postoperative osteomyelitis: Fabrication, characterization, and in vitro–in vivo evaluation. Carbohydrate Polymers. 2020;**244**:116482

[5] Alshaaer M, Kailani MH, Jafar H, Ababneh N, Awidi A. Physicochemical and microstructural characterization of injectable load-bearing calcium phosphate scaffold. Advances in Materials Science and Engineering. 2013;**2013**:8. Article ID: 149261

[6] Khalifehzadeh R, Arami H. Biodegradable calcium phosphate nanoparticles for cancer therapy. Advances in Colloid and Interface Science. 2020;**279**:102157. DOI: 10.1016/j.cis.2020.102157

[7] Shyong Y-J, Chang K-C, Lin F-H. Calcium phosphate particles stimulate exosome secretion from phagocytes for the enhancement of drug delivery. Colloids and Surfaces B: Biointerfaces. 2018;**1711**:391-397

[8] Alshaaer M, Cuypers H, Rahier H, Wastiels J. Production of monetite-based Inorganic phosphate cement (M-IPC) using hydrothermal post curing (HTPC). Cement and Concrete Research. 2011;**41**:30-37

[9] Alshaaer M, Cuypers H, Mosselmans G, Rahier H, Wastiels J. Evaluation of a low temperature hardening inorganic phosphate cement for high-temperature applications. Cement and Concrete Research. 2011;**41**:38-45

[10] Zhang J, Liu W, Schnitzler V, Tancret F, Bouler J-M. Calcium phosphate cements for bone substitution: Chemistry, handling and mechanical properties. Acta Biomaterialia. 2014;**10**:1035-1049

[11] Mert I, Mandel S, Tas AC. Do cell culture solutions transform brushite (CaHP04 2H20) to octacalium phosphate (Ca8(HP04)2(P04)4 5H20)? In: Narayan R, Colombo P, editors. Advances in Bioceramics and Porous Ceramics IV. Hoboken, New Jersey: John Wiley & Sons, Inc; 2011. pp. 79-94

[12] Hurle K, Oliveira J, Reis R, Pina S, Goetz-Neunhoeffer F. Ion-doped brushite cements for bone regeneration. Acta Biomaterialia. 2021;**123**:51-71

[13] Dosen A, Giese RF. Thermal decomposition of brushite, CaHPO4·2H2O to monetite CaHPO4 and the formation of an amorphous phase. American Mineralogist. 2011;**96**(2-3):368-373

[14] Patil SB, Jena A, Bhargava P. Influence of ethanol amount during washing on deagglomeration of co-precipitated calcined nanocrystalline 3YSZ powders. International Journal of Applied Ceramic Technology. 2012. DOI: 10.1111/j.1744-7402.2012.02813.x

[15] Piva RH, Piva DH, Pierri J, Montedo ORK, Morelli MR. Azeotropic

distillation, ethanol washing, and freeze drying on coprecipitated gels for production of high surface area 3Y–TZP and 8YSZ powders: A comparative study. Ceramics International. 2015;**41**:14148-14156

[16] Alshaaer M, Afify A, Moustapha M, Hamad N, Hammouda G, Rocha F. Effects of the full-scale substitution of strontium for calcium on the microstructure of brushite: (CaxSr1–x)HPO4.nH2O system. Clay Minerals. 2020;**55**(4):366-374

[17] Alshaaer M. Microstructural characteristics and long-term stability of wollastonite-based chemically bonded phosphate ceramics. International Journal of Applied Ceramic Technology. 2021;**18**:319-331

[18] Bhojani A, Jethva H, Joshi M. Growth inhibition study of urinary type brushite crystal using potassium dihydrogen citrate solution. AIP Conference Proceedings. 2019;**2115**:030417. DOI: 10.1063/1.5113256

[19] Wu X, Gu J. Inorganic resins composition, their preparation and use thereof. Belgium/Brussels Patent EP 0 861 216 B1. 2000

[20] Mosselmans G, Biesemans M, Willem R, Wastiels J, Leermakers M, Rahier H, et al. Thermal hardening and structure of a phosphorus containing cementitious model material. Journal of Thermal Analysis and Calorimetry. 2007;**88**:723-729

[21] Sayahi M, Santos J, El-Feki H, Charvillat C, Bosc F, Karacan I, et al. Brushite (Ca,M)HPO4, 2H2O doping with bioactive ions (M = Mg2+, Sr2+, Zn2+, Cu2+, and Ag+): A new path to functional biomaterials? Materials Today Chemistry. 2020;**16**:100230. DOI: 10.1016/j.mtchem.2019.100230

[22] G. Girişken and A. C. Taş, Development of biomineralization solutions to facilitate the transformation of brushite (CaHPO4·2H2O) into octacalcium phosphate (Ca8(HPO4) 2(PO4)4·5H2O), In: 15th National Biomedical Engineering Meeting (BIYOMUT), Antalya, Turkey; 2020

[23] Xue Z, Wang Z, Sun A, Huang J, Wu W, Chen M, et al. Rapid construction of polyetheretherketone (PEEK) biological implants incorporated with brushite (CaHPO4·2H2O) and antibiotics for anti-infection and enhanced osseointegration. Materials Science and Engineering: C. 2020;**111**:110782. DOI: 10.1016/j.msec.2020.110782

[24] Alshaaer M, Cuypers H, Rahier H, Wastiels J. Production of monetite-based inorganic phosphate cement (M-IPC) using. Cement and Concrete Research. 2011;**41**:30-37

[25] Frost RL, Palmer SJ. Thermal stability of the 'cave' mineral brushite CaHPO4·2H2O—Mechanism of formation and decomposition. Thermochimica Acta. 2011;**521**(1-2):14-17

[26] Tamimi F, Le Nihouannen D, Eimar H, Sheikh Z, Komarova S, Barralet J. The effect o fautoclaving on the physical and biological properties of dicalcium phosphate dihydrate bioceramics: Brushite vs. monetite. Acta Biomaterialia. 2012;**8**(8):3161-3169

[27] Tortet L, Gavarri JR, Nihoul G. Study of protonic mobility in CaHPO4 2H2O (Brushite) and CaHPO4 (Monetite) by infrared spectroscopy and neutron scattering. Journal of Solid State Chemistry. 1997;**132**:6-16

[28] Luo J, Engqvist H, Persson C. A ready-to-use acidic, brushite-forming calcium phosphate cement. Acta Biomaterialia. 2018;**81**:304-314

[29] Han L, Qi-Zhi Y, Yu-Ying W, Yi-Liang L, Gen-Tao Z. Biomimetic synthesis of struvite with biogenic morphology and implication for

pathological biomineralization. Scientific Reports. 2015;5:7718. DOI: 10.1038/srep07718

[30] Ding H, Pan H, Xu X, Tang R. Toward a detailed understanding of magnesium ions on hydroxyapatite crystallization inhibition. Crystal Growth & Design. 2014;**14**(2):763-769

[31] Li H, Yao Q-Z, Wang Y-Y, Li Y-L, Zhou G-T. Biomimetic synthesis of struvite with biogenic morphology and implication for pathological biomineralization. Scientific Reports. 2015;**5**(1):7718

[32] Nosrati H, Le DQS, Zolfaghari RE, Canillas MP, Bünger CE. Nucleation and growth of brushite crystals on the graphene sheets applicable in bone cement. Boletín de la Sociedad Española de Cerámica y Vidrio. 2020. DOI: 10.1016/j.bsecv.2020.05.001

[33] Alshaaer M, Abdel-Fattah E, Saadeddin I, Al Battah F, Issa KI, Saffarini G. The effect of natural fibres template on the chemical and structural properties of Biphasic Calcium Phosphate scaffold. Materials Research Express. 2020;7(6):065405. DOI: 10.1088/2053-1591/ab9993

Chapter 8

Advanced Biocrystallogenesis

Ivana Kuta Smatanova, Petra Havlickova, Barbora Kascakova and Tatyana Prudnikova

Abstract

Nowadays, X-ray crystallography is one of the most popular structural biology methods. Successful crystallization depends not only on the quality of the protein sample, precipitant composition, pH or other biophysical and biochemical parameters, but also largely on the use of crystallization technique. Some proteins are difficult to be crystallized using basic crystallization methods; therefore, several advanced methods for macromolecular crystallization have been developed. This chapter briefly reviews the most promising advanced crystallization techniques and strategies as one of the efficient tools for crystallization of macromolecules. Crystallization in capillaries, gels, microfluidic chips, electric and magnetic fields as well as crystallization under microgravity condition and crystallization in living cells are briefly described.

Keywords: protein crystallization, protein crystal, advanced methods, crystallization strategies, biocrystallogenesis

1. Introduction

Macromolecular crystallization was invented accidentally in the late 19th century. The initial goal of crystallization processes was to purify and to prove the chemical purity of the examined chemical compound. One of the first achievements in macromolecular crystallization were crystals of hen-egg albumin at the end of the 19th century and crystals of insulin in the 1920s [1, 2]. Since then, macromolecular crystallization has developed into a powerful tool for the three-dimensional structure determination of nucleic acids, proteins as well as for larger macromolecular complexes.

A good-quality macromolecular crystal is needed in X-ray crystallography for obtaining a 3D model of corresponding nucleic acids, proteins, macromolecular complexes or larger biological assemblies, for instance, viruses or ribosomes. Nevertheless, macromolecular crystallization is a highly unpredictable process, which relies on a large number of chemical and biochemical parameters, such as the purity and concentration of the protein, type of precipitant, pH of the buffer, as well as on physical parameters, as temperature, time, pressure or vibrations, etc.

In order to crystallize the protein from the solution, the supersaturated state has to be reached using various crystallization techniques. Crystallization process can be divided into three stages - nucleation, crystal growth and termination. So-called phase diagrams typically illustrate these steps. For the investigation of the best conditions for crystal growth of the individual macromolecule, two general strategies are usually applied. Firstly, initial crystallization screening using commercially available screening kits is performed to explore the suitable conditions for crystal growth. Secondly, conditions yielding macromolecular crystals are systematically modified to allow the growth of the best-quality crystals adequate for X-ray diffraction analysis. Basic crystallization techniques, namely batch, vapour diffusion and free-interface diffusion are used for initial crystallization screening generally [3]. In addition, these techniques can be further modified to obtain as good-quality crystals as possible during the optimization of protein crystallization. However, some proteins cannot be easily crystallized and produced crystals in diffraction quality and thus initial hits need to be further improved. It is important to note that different crystallization methods screen the phase diagram from different points of view and hence affect the properties of the resulting crystals. Therefore, advanced biocrystallogenesis methods are recommended to be applied for screening or optimization of crystallization conditions.

This chapter reviews different advanced methods and strategies as efficient tools for crystallization of macromolecules. During the past few years, the field of protein crystallography has significantly developed regarding instrumentation. Different technologies require different sizes of crystals, for instance homogeneous nanocrystals for XFEL or big size crystals for neutron diffraction [4, 5]. This also drives the necessity to develop new and convenient methods for growing suitable high-quality protein crystals. For this purpose advanced crystallization methods and strategies have been developed.

2. Advanced biocrystallogenesis methods

2.1 Counter-diffusion

Counter-diffusion belongs to the non-equilibrium crystallization approaches. The principle of this method is based on the use of one-dimensional experimental chambers, including protein and precipitant solutions, resulting in slow mutual diffusion, in other words the slow mixing of the molecules of the solutions and therefore slow reaching of the supersaturation state [6]. During the diffusion, the solubility of the macromolecule is significantly reduced (**Figure 1**–(3)) [7].

The ideal concentration of the protein solution is advised to be higher (more than 4 mg/ml) and the concentration of the precipitant solution should be very high, nearly reaching the supersaturation state [8]. At the beginning of the experiment, the diffusion is faster and thus the amorphous precipitate is formed. Eventually, the protein concentration decreases at the liquid-liquid interface, which leads to the nucleation followed by the protein crystal growth [9].

Counter-diffusion method for protein crystallization and screening is popular since it screens a wide range of the phase diagram while carrying out only one experiment and increases the chance of hits at the same time.

2.2 Crystallization in capillaries

Crystallization in thin and long capillaries is one of the most used advanced crystallization technique, frequently applying counter-diffusion. Furthermore, capillaries can be used also in batch crystallization or vapour diffusion (**Figure 1**–(2)). This technique does not require further optimization as it precisely screens many crystallization conditions for one protein solution and one precipitant solution. Another advantage of thin capillaries is that a small amount of the protein solution is required for the experiment. Moreover, the protein crystals in capillaries can be also directly measured at the synchrotrons [10].

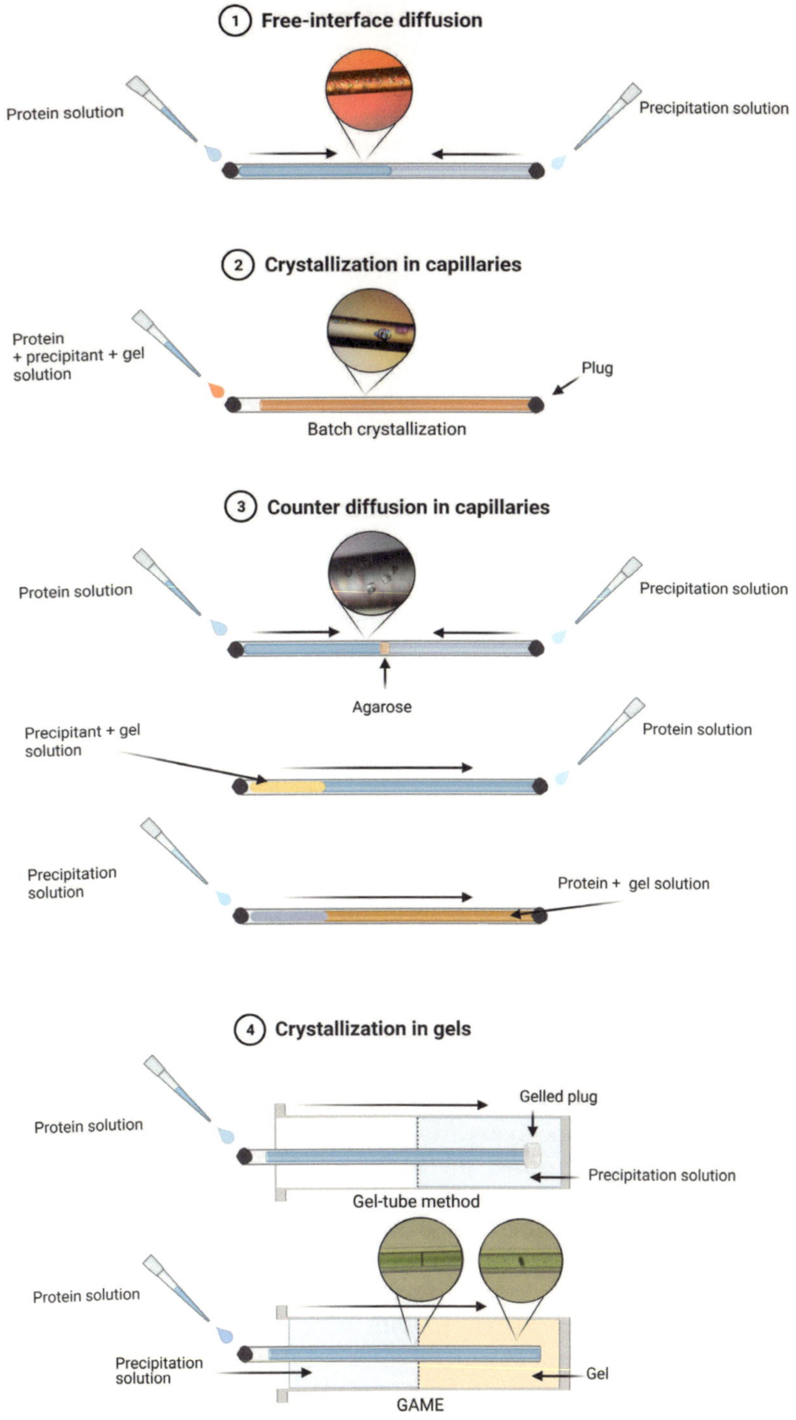

Figure 1.
Advanced biocrystallogenesis methods. 1. Free-interface diffusion, 2. Crystallization in capillaries, 3. Counter-diffusion in capillaries, 4. Crystallization in gels and GAME (prepared by author Barbora Kascakova).

Three different organizations of capillaries can be applied for counter-diffusion experiments: (i) simple one chamber organization composed of the protein and precipitant solutions, (ii) two chamber single capillary organization is formed by the precipitant solution and the protein solution mixed with the gel and (iii) three chamber single capillary organisation that blocks fast diffusion of the protein and precipitant solutions by a physical barrier, typically formed by agarose or silica gel, known as gel acupuncture method (GAME) (**Figure 1**–(1, 3, 4)). Modifications (ii) and (iii) have been successfully used for the crystallization of not only globular proteins, but also RNA and DNA protein complexes and larger membrane protein complexes [10–12].

2.3 Crystallization in gels

Crystallization in gels is a suitable method for growth of high-quality macromolecular crystals. This method is effective due to the invariable environment enabling crystal growth, preventing the production of protein aggregates or sedimentation and slows diffusion of molecules in the crystallization chamber [13]. The most used gels for macromolecular crystallization are hydrophilic gels, namely agarose and silica gels, even though they have different properties triggering the crystal growth [14].

Regarding nucleation, agarose gels significantly promote it, whereas, silica gels inhibit this process, but the exact mechanism behind it remains unknown. The benefits of the use of agarose gels are (i) the concentration of agarose gel can be very low, (ii) agarose gel polymerises faster than silica gels, (iii) agarose gel can be employed in any crystallization technique, (iv) cofactors, ligands and even cryoprotectants are reported to be easily soaked into the crystal in the presence of the agarose gel, (v) agarose gels help to produce bigger crystals for neutron diffraction analysis and also (vi) serves as delivery medium for nanocrystals for X-ray free-electron laser analysis (**Figure 1**–(3, 4)) [14–17].

2.4 Microgravity

The first attempts to confirm the substantial role of microgravity for crystal growth were done more than 35 years ago [18]. There was an assumption that absence of gravity can lead to growing of top quality crystals, thus crystallization experiments in the microgravity environment in the space shuttle were carried out. First experiments were performed by vapour diffusion method, later the counter diffusion and dialysis crystallization were applied as well [18, 19]. The main advantage of growing protein crystals in space is that there is very diffusive mass transport and reduction of formation of possible crystal defects.

There are different views on the success of this method. In many cases, it was proven that space provides good conditions for growing better diffractive crystals against the results of crystals grown on Earth [20, 21]. Many positive examples of growing the crystals in space show improved diffraction resolution, better diffraction intensity, reduction of crystal disordering and clustering, prolonged nucleation process, excluded crystal sedimentation and in some cases growing of larger crystals, sometimes also with different space groups [20, 22]. Of course, the microgravity experiments are connected with some limitations that are mainly bureaucracy for flight approval of protein samples, delays of mission and with this connected possible degradation of protein or crystal samples [22].

2.5 Microfluidic chips

Recently, screening and crystallization by counter diffusion method has been improved by the application of microfluidic chips. Typically, the chips are composed of channels connected by a cross section. This organization of the crystallization chamber facilitates the screening of the wide range of the protein and precipitant concentrations; the supersaturation state is achieved by diffusion of the molecules [9]. Another advantage of this approach is the requirement of a small amount of the protein and thus is suitable for proteins that are demanding to purify in larger volumes [23]. In practice, the protein sample is loaded into the eight channels at once by micropipettes and precipitant solutions are placed in the reservoirs [24]. The crystal growth in microfluidic chips is much faster than in other crystallization systems and the results are highly reproducible [25]. Microfluidic chips are also applicable for the microseeding strategy or soaking with the ligands [26]. The resulting crystals can be analysed by *in situ* serial crystallography [24, 27].

2.6 Electric and magnetic field

The principle of the method benefiting from electric fields relies on the separation of the nucleation and crystal growth; it can be applied in the electric as well as the electromagnetic field. The nucleation process takes place while the current is switched off [28]. In order to perform crystallization experiments in the electric field, the specially designed crystal growth cell has to be used. The system contains two parallel electrodes and is connected to a direct current source that provides direct or alternating current. Nevertheless, the experiment has to be set up before the application of the current. It was reported that the current positively influences the size of protein crystals. Specifically, lower frequencies of alternating current provide less larger crystals compared to higher frequencies [28]. For the purposes of using electric current the e-crystallization cells based on vapour diffusion or batch method were made [29].

In the case of using magnetic fields, there is a similar outcome as carrying out an experiment in microgravity. This can be achieved when using a vertical magnetic field with use of anti-gravitational force. The crystals treated by magnetic fields show better diffraction quality as was revealed for microgravity experiments [30, 31]. In addition, the crystallization conditions need to be known before initiating the experiment. Moreno 2017 mentions the use of gels for obtaining large well-arranged crystals together with use of magnetic forces for at best two days and more days. For this type of experiment, the capillary glass pipettes were reported to be suitable for use of homogeneous or nonhomogeneous magnetic fields [28]. The best influence of magnetic field use is regulation of quality and crystal size [32].

3. Advanced biocrystallogenesis strategies

3.1 Seeding

The supersaturation level needed for the nucleation and the growth of best diffracting crystals is often different. Nucleation takes place when the supersaturation level is very high, while crystal-growing process appears when the conditions are slightly changing to lower supersaturation level [33]. Therefore, the seeding strategy has been developed as one of the most powerful tools for the optimization of the size and diffractive quality of single crystal. Another benefit is that it reduces the amount of the used protein and saves time needed for the spontaneous nucleation. This is also suitable for the examination of the slightly changing crystallization conditions, such as testing different pH ranges, adding new additives or even new precipitants. Several seeding strategies such as macroseeding, microseeding and/or their modifications can be applied.

In order to perform macroseeding, a single seed crystal has to be selected; its size is not relevant. The solubility of the crystal has to be investigated by testing of different precipitant concentrations to establish the conditions where the crystal is stable. This method is more demanding as multiple washes of the crystal are needed before the crystal is put into the suitable equilibration condition. The washes are aimed to eliminate additional nuclei. The main disadvantage of the method is that during this handling, the crystal seeds can be damaged and can form only the microcrystals in the new equilibration condition, which are not suitable for the X-ray diffraction analysis [33].

Microseeding can be described as a delivery of microseeds into a new equilibration condition. Before the experiment, the seed stock has to be prepared by crushing crystals using vortex, sonication or seed beads [34]. Once the seed stock is prepared, the dilution series have to be done in order to stabilize the nuclei presented in the seed stock and enable them to form suitable crystals in the new crystallization drop.

Another way of seeding strategy is the streak seeding. The drop containing the protein crystal is opened and the crystal is touched using an animal (cat, horse, rabbit or chinchilla) whisker. The nuclei can be attached to the whisker and transfer into a new crystallization drop. The sitting drop technique is recommended for this experiment, as it does not dry as quickly as the drops in hanging drop technique [33].

Cross seeding is specifically designed for crystallization of related proteins. These related proteins are different recombinantly modified protein variants, homologous proteins, and chemically modified proteins, macromolecular complexes with ligand or even misfolded proteins that are crystallizing difficult [35]. Some of the related proteins do not crystallize in the same conditions; however, sometimes this method can help and save the protein solution and time.

Random microseeding is a modified microseeding strategy using random screening kits. It is reported to create extra additional hits as well as produce better quality protein crystals for diffraction experiments [36].

3.2 Co-crystallization with ligands and crystallization chaperones

The co-crystallization of macromolecules with ligands have been developed because some proteins crystallize efficiently with ligands considering its structure stabilization role. The use of ligands has benefits not only for improving crystallization but also for obtaining stable and functional protein during purification because they enhance thermal stability of molecules. The study of small ligands stabilization roles for proteins in aqueous solution is still in progress. There exist online libraries of ligands as compounds used for crystallization purposes to prepare the best condition for protein stabilization [37]. The use of ligands for crystallization promotes the possibility of obtaining good-quality protein crystals. Ligands are regularly used to expand the screening capacity of commercially available screens [38]. On the contrary, it can be challenging to crystalize protein-ligand complex and so, the use of stabilizing additives is necessary. The process of co-crystallization can be further affected by temperature used for the complex formation prior its crystallization and by concentration of protein or used ligand [39].

Another "compounds" used for stabilization of proteins are amino acids and their derivatives such as glycine ethyl ester or glycine amide. There is noted considerable benefit of the stabilization role of amino acids for protein purification as well as crystallization [40, 41]. Hence, small molecule additives such as coenzymes, prosthetic groups, inhibitors and small molecules, mentioned in [42], are used for successful crystallization of macromolecules. Usage of these additives helps forming hydrogen bonds and crosslinks between molecules that helps formation of stable crystals with regular lattice order of molecules [38].

Additionally, the antibodies and nanobodies are used to improve protein crystallization. Antibodies such as conventional IgG, Fab-fragments (fragment antigen binding), scFv (single-chain variable fragments) and Camelid hcAb (heavy chain IgG) were used to reinforce protein crystallization, mainly for membrane protein co-crystallization. Latterly, nanobodies replaced the use of antibodies because production of antibodies is limited and obtained yield tends to be conformationally heterogeneous with limited solubility [43, 44]. On the contrary, nanobodies are extremely stable and soluble antibody single-domain fragments that stabilize unstructured proteins by binding to them, form crystal contacts and thus speed up crystal growth. For crystallization of selected protein, many conformationally different nanobodies can be used to enhance the probability of best quality crystal hits [45].

3.3 Protein engineering

Protein engineering is one of the most used strategies in molecular biology to produce sufficient amounts of soluble protein. Protein biocrystallogenesis is likewise using protein engineering for crystallization of proteins that are hard to crystallize owing to have highly flexible regions, high surface entropy or less common high surface hydrophobicity or proteins are not able to be produced soluble [46]. To conquer these issues some strategies such as designing modified proteins, mutation of surface residues or designing fusion proteins are used [47].

The modifications of proteins include designing only some parts of protein such as conserved domains, and removing flexible parts or internal loops that prevent crystal formation. This can be fulfilled by creation of genetic constructs based on e.g. computation selection of expression clone libraries to find the clones with fragments that would encode the concerned part of the protein [47]. In the literature, several strategies for protein modification have been mentioned such as colony filtration blot technique [48], the open reading frame selection through the ESPRIT automation to find soluble complexes [49] and fluorescence screening based on protein fusion with GFP protein [47, 50].

Next strategy to enhance crystal formation is design of protein mutants by site-specific mutation or chemical modification. Mutagenesis is ordinarily used to produce soluble protein assuming that wild type is not soluble. The substitution of Lys, Gln or Glu that are frequently located on the molecule surface [51] with Ala that has lower surface entropy is a commonly used strategy [52]. There are plenty of protein structures that were solved mainly due to mutagenesis [47]. In addition, many chemical modifications as reductive methylation of Lys that reduce protein solubility and reductive carboxymethylation or substitution of Cys that on the contrary enhance solubility, stability and have benefits for protein function and oligomerization [52]. Different strategy is to introduce Cys at the surface to strengthen formation of symmetrical dimers that seem to crystallize more easily [52, 53].

Design of fusion proteins or fusion tags is another approach to produce soluble proteins and subsequently solve protein structures. In many cases the fusion partner such as short *Strep*-tag or poly-histidine tag and some larger tags as glutathione S-transferase (GST) and Maltose-binding protein (MBP) are removed before crystallization [54–56]. This is done to remove possible highly flexible regions that can block formation of protein contacts during crystallization. However, it was shown that in two different approaches fusion proteins manage to crystalize target proteins. First, one is based on increasing protein surface area by incorporation of fused protein such as MBP, Green fluorescent protein (GFP), Barnase or T4 lysozyme (T4L) [56]. Second is aimed to merge interacting proteins or protein and peptide with linker, mostly to stabilize one of fused proteins or to support their interaction [56, 57].

3.4 Crystallization in living cells

The formation of native protein crystals in living cells has been detected over the last few decades as a natural process. This approach is ideal for proteins that cannot be crystallized applying basic crystallization techniques. Recently, X-ray sources have been exceedingly improved, such as microfocus beamlines, X-ray free-electron lasers, or even serial crystallography as a new data collection strategy [58, 59]. These advances in the field of X-ray crystallography enable data collection from smaller crystals and thus enable the development of *in cellulo* crystallization as a powerful advanced crystallization approach.

Recombinant or fusion recombinant proteins can be crystallized in a number of living cells, including plant cells, mammalian cells or insect cells [60]. The bottleneck of this strategy is the detection of the protein crystals inside the cells. Crystals can occur in different compartments of the cell, such as cytosol, endoplasmic reticulum (ER) or peroxisomes [60, 61]. The crystals can be detected using various methods, e.g. bright-field microscopy techniques or transmission electron microscopy (TEM), which is also used for the observation of crystal growth and subsequent optimization. The crystal can be isolated from the cell before the diffraction experiment or directly measured in the cell [60, 62].

3.5 Automation

The expansion of the crystallographic field and running extensive research led to need for facilitation of performing crystallization experiments. This was achieved by automation that supported faster progress of experiments thanks to development of a variety of robots [63]. There are robots adjusted to establishment of crystallization conditions, arrangement of drops, for carrying out the vapour diffusion experiments specifically hanging drop, microbatch, free-interface diffusion and random micro-seeding seeding methods with sitting drop method as the most preferred method. Most widely used crystallization robots are from the Oryx series (Douglas Instruments Ltd.), the Mosquito (TTP Labtech, Royston, UK) and TOPAZ system (Fluidigm Corp., San Francisco CA, USA) [64]. Automation devices allow crystallization of nanovolumes into an array of plates for 96 or more conditions [28]. Each robot has its own specification, from which can be chosen preferred experimental method suitable for protein of interest.

4. Conclusion

In conclusion, a growth the good-quality macromolecular crystal is crucial for the determination of its atomic structure using of X-ray crystallography. Several effective advanced methods leading to the formation of crystal in diffraction quality are available. Nowadays, the extensive improvement of X-ray facilities enables data collection from smaller crystals and allows using the new trends in crystallization strategies that are developing as a powerful advanced crystallization approach.

The protein crystallization manufacturers such as Hampton Research, Molecular Dimensions Ltd., PerkinElmer, GE Healthcare, Danaher, Bruker, Agilent, Jena Bioscience, Rigaku, Formulatrix and MiTeGen are basic companies responsible for systematic invention of new products and technologies that constantly improve protein crystallization methods and are looking for crystallization techniques for hardly crystallisable proteins such as membrane proteins.

Next essentials that have an effort to improve crystallization techniques are attempts to use crystals in different non-conventional ways. One of them is sending biosubstances (protein-based medicines) into space to try to crystallize them for finding new uses of crystallized proteins as pharmacotherapeutics. These protein crystals can be supposedly thereafter used for long-term storage or for stabilization of biosubstances by transferring them into solid state (Bristol-Myers

Squibb Company). Another use of crystallization was reported in material sciences, more precisely for improvement of electronic materials developed on the base of biomolecules arrangement in crystals [65].

These and many more applications of protein crystallization are reasons for development of new methods and strategies for faster crystallization and obtaining of well diffractive crystals and for reduction of protein volume. However, there is still a necessity for improvement and new methods and strategies for biomolecules crystallization awaits to be discovered. Although this can be achieved more easily because biochemistry, biotechnology, physics, chemistry and applied nanosciences are extra overlapped and allow to improve and explain in depth the process of protein crystallization.

Acknowledgements

This research was supported by ERDF (grant No. CZ.02.1.01/0.0/0.0/15_003/0000441 and GAJU 106/2021/P).

Conflict of interest

The authors declare that the research was conducted in the absence of any commercial or financial relationships that could be construed as a potential conflict of interest.

Author details

Ivana Kuta Smatanova*, Petra Havlickova†, Barbora Kascakova†
and Tatyana Prudnikova
Faculty of Science, Department of Chemistry, University of South Bohemia in Ceske Budejovice, Ceske Budejovice, Czech Republic

*Address all correspondence to: ivanaks@seznam.cz

† These authors have contributed equally.

References

[1] Hofmeister, T. (1890). *Z. Physiol. Chem.* **14**, 165.

[2] Abel, J. (1927). Chemistry in Relation to Biology and Medicine with Especial Reference to Insulin and Other Hormones. *Science, 66*(1710), 307-319. Retrieved February 22, 2021, from http://www.jstor.org/stable/1651597

[3] McPherson A, Gavira JA. Introduction to protein crystallization. Acta Crystallogr F Struct Biol Commun. 2014 Jan;70(Pt 1):2-20. doi: 10.1107/S2053230X13033141. Epub 2013 Dec 24. PMID: 24419610; PMCID: PMC3943105.

[4] Boudes M, Garriga D, Coulibaly FJ. Reflections on the Many Facets of Protein Microcrystallography. Australian Journal of Chemistry. 2014;67(12):1793-1806. https://doi.org/10.1071/CH14455

[5] Nannenga BL, Gonen T. Protein structure determination by MicroED. Curr Opin Struct Biol. 2014 Aug;27:24-31. doi: 10.1016/j.sbi.2014.03.004. Epub 2014 Apr 5. PMID: 24709395; PMCID: PMC5656570.

[6] Manuel García-Ruiz J. Nucleation of protein crystals. J Struct Biol. 2003 Apr;142(1):22-31. doi: 10.1016/s1047-8477(03)00035-2. PMID: 12718916.

[7] Otálora F, Gavira JA,Ng JD, García-Ruiza JM. Counter diffusion methods applied to protein crystallization. Progress in Biophysics and Molecular Biology. 2009;101(1-3):26-37. DOI: 10.1016/j.pbiomolbio.2009.12.004.

[8] Henisch HK, García-Ruiz JM. Crystal growth in gels and Liesegang ring formation: I. Diffusion relationships. Journal of Crystal Growth. 1986;75(2): 195-202. https://doi.org/10.1016/0022-0248(86)90028-X.

[9] García-Ruiza JM, Novella ML, Moreno R, Gavira JA. Agarose as crystallization media for proteins: I: Transport processes. Journal of Crystal Growth. 2001;232(1-4):165-172. ISSN 0022-0248. https://doi.org/10.1016/S0022-0248(01)01146-0.

[10] López-Jaramillo FJ, García-Ruiz JM, Gavira JA and F. Otálora. Crystallization and cryocrystallography inside X-ray capillaries. J. Appl. Cryst. 2001;34:365-370. https://doi.org/10.1107/S0021889801003697

[11] Biertümpfel C, Basquin J, Suck D, Sauter C. Crystallization of biological macromolecules using agarose gel. Acta Crystallogr D Biol Crystallogr. 2002 Oct;58(Pt 10 Pt 1):1657-9. doi: 10.1107/s0907444902012738. Epub 2002 Sep 26. PMID: 12351881.

[12] Kutá Smatanová, I., Gavira, J.A., Řezáčová, P. *et al.* New techniques for membrane protein crystallization tested on photosystem II core complex of *Pisum sativum* . *Photosynth Res* 90, 255-259 (2006). https://doi.org/10.1007/s11120-007-9131-y

[13] Gavira JA, de Jesus W, Camara-Artigas A, López-Garriga J, García-Ruiz JM. Capillary crystallization and molecular-replacement solution of haemoglobin II from the clam Lucina pectinata. Acta Crystallogr Sect F Struct Biol Cryst Commun. 2006 Mar 1;62(Pt 3):196-9. doi: 10.1107/S1744309106002648. Epub 2006 Feb 10. PMID: 16511300; PMCID: PMC2197196.

[14] Artusio F, Castellví A, Sacristán A, Pisano R, and JA Gavira. Agarose Gel as a Medium for Growing and Tailoring Protein Crystals. Cryst. Growth Des. 2020;20(8):5564-5571. https://doi.org/10.1021/acs.cgd.0c00736

[15] Gavira JA, García-Ruiz JM. Agarose as crystallisation media for proteins II: trapping of gel fibres into the crystals. Acta Crystallogr D Biol Crystallogr.

2002 Oct;58(Pt 10 Pt 1):1653-6. doi: 10.1107/s0907444902014609. Epub 2002 Sep 26. PMID: 12351880.

[16] Sugiyama Y, Ikarugi S, Oura K, Ikeda A, Matsuyama E, Ono R, Nomura M, Tawarayama H, Saito T, Kuwahara K. MFI zeolite membranes prepared on novel silica substrates. Journal of Chemical Engineering of Japan. 2015 ;48(11):891-896. https://doi.org/10.1252/jcej.15we014

[17] Conrad CE, Basu S, James D, Wang D, Schaffer A, Roy-Chowdhury S, Zatsepin NA, Aquila A, Coe J, Gati C, Hunter MS, Koglin JE, Kupitz C, Nelson G, Subramanian G, White TA, Zhao Y, Zook J, Boutet S, Cherezov V, Spence JC, Fromme R, Weierstall U, Fromme P. A novel inert crystal delivery medium for serial femtosecond crystallography. IUCrJ. 2015 Jun 30;2(Pt 4):421-30. doi: 10.1107/S2052252515009811. PMID: 26177184; PMCID: PMC4491314.

[18] Littke W, John C. Materials: protein single crystal growth under micro-gravity. Science. 1984 Jul 13;225(4658): 203-4. doi: 10.1126/science.225.4658.203. PMID: 17837944.

[19] DeLucas LJ, Suddath FL, Snyder R, Naumann R, Broom MB, Pusey M, Yost V, Herren B, Carter D, Nelson B, Meehan EJ, McPherson A, Bugg ChE. Preliminary investigations of protein crystal growth using the space shuttle. Journal of Crystal Growth. 1986;76(3):681-693. ISSN 0022-0248, https://doi.org/10.1016/0022-0248(86)90185-5.

[20] Ng JD. Space-grown protein crystals are more useful for structure determination. Ann N Y Acad Sci. 2002 Oct;974:598-609. doi: 10.1111/j.1749-6632.2002.tb05934.x. PMID: 12446351.

[21] Vergara, A., Lorber, B., Sauter, C., Giegé, R., & Zagari, A. (2005). Lessons from crystals grown in the Advanced Protein Crystallisation Facility for conventional crystallisation applied to structural biology. Biophysical chemistry, 118(2-3), 102-112. https://doi.org/10.1016/j.bpc.2005.06.014

[22] McPherson A, DeLucas LJ. Microgravity protein crystallization. NPJ Microgravity. 2015 Sep 3;1:15010. doi: 10.1038/npjmgrav.2015.10. PMID: 28725714; PMCID: PMC5515504.

[23] van der Woerd M, Ferree D, Pusey M. The promise of macro-molecular crystallization in microfluidic chips. J Struct Biol. 2003 Apr;142(1): 180-7. doi: 10.1016/s1047-8477(03)00049-2. PMID: 12718930.

[24] de Wijn R, Hennig O, Roche J, Engilberge S, Rollet K, Fernandez-Millan P, Brillet K, Betat H, Mörl M, Roussel A, Girard E, Mueller-Dieckmann C, Fox GC, Olieric V, Gavira JA, Lorber B, Sauter C. A simple and versatile microfluidic device for efficient biomacromolecule crystallization and structural analysis by serial crystallography. IUCrJ. 2019 Apr 19;6(Pt 3):454-464. doi: 10.1107/S2052252519003622. PMID: 31098026; PMCID: PMC6503916.

[25] Abdalla M, Ali Eltayb W, Samad A, Elshareef SHM, Dafaalla TIM. Important Factors Influencing Protein Crystallization. Glob J Biotechnol Biomater Sci. 2016;2(1):025-028 DOI: 10.17352/gjbbs.000008

[26] Martin-Garcia JM, Conrad CE, Coe J, Roy-Chowdhury S, Fromme P. Serial femtosecond crystallography: A revolution in structural biology. Arch Biochem Biophys. 2016 Jul 15;602:32-47. doi: 10.1016/j.abb.2016.03.036. Epub 2016 Apr 30. PMID: 27143509; PMCID: PMC4909539.

[27] Johansson LC, Stauch B, Ishchenko A, Cherezov V. A Bright Future for Serial Femtosecond Crystallography with XFELs. Trends

Biochem Sci. 2017 Sep;42(9):749-762. doi: 10.1016/j.tibs.2017.06.007. Epub 2017 Jul 18. PMID: 28733116; PMCID: PMC5600683.

[28] Moreno A. Advanced Methods of Protein Crystallization. Methods Mol Biol. 2017;1607:51-76. doi: 10.1007/978-1-4939-7000-1_3. PMID: 28573569.

[29] Flores-Hernández E, Stojanoff V, Arreguín-Espinosa R, Moreno A, Sánchez-Puig N. An electrically assisted device for protein crystallization in a vapor-diffusion setup. J Appl Crystallogr. 2013 Jun 1;46(Pt 3):832-834. doi: 10.1107/S0021889813010558. Epub 2013 May 15. PMID: 23682197; PMCID: PMC3654317.

[30] Wakayama NI, Yin DC, Harata K, Kiyoshi T, Fujiwara M, Tanimoto Y. Macromolecular crystallization in microgravity generated by a superconducting magnet. Ann N Y Acad Sci. 2006 Sep;1077:184-93. doi: 10.1196/annals.1362.024. PMID: 17124123.

[31] Wada H, Hirota N, Matsumoto S, Okada H, Kiyohara M, Ode T, Tanokura M, Nakamura A, Ohtsuka J, Kita A, Numoto N, Kashiwagi T, Suzuki E. Application of High-Field Superconducting Magnet to Protein Crystallization. Physics Procedia. 2012;36:953-957. ISSN 1875-3892. https://doi.org/10.1016/j.phpro.2012.06.236.

[32] Sazaki G, Moreno A, Nakajima K. Novel coupling effects of the magnetic and electric fields on protein crystallization. Journal of Crystal Growth. 2004; 262(1-4):499-502. ISSN 0022-0248. https://doi.org/10.1016/j.jcrysgro.2003.09.050.

[33] Bergfors T. Seeds to crystals. J Struct Biol. 2003 Apr;142(1):66-76. doi: 10.1016/s1047-8477(03)00039-x. PMID: 12718920.

[34] Luft JR, DeTitta GT. A method to produce microseed stock for use in the crystallization of biological macromolecules. Acta Crystallogr D Biol Crystallogr. 1999 May;55(Pt 5): 988-93. doi: 10.1107/s090744499 9002085. PMID: 10216295.

[35] Abuhammad A, McDonough MA, Brem J, Makena A, Johnson S, Schofield ChJ, and EF Garman. "To Cross-Seed or Not To Cross-Seed": A Pilot Study Using Metallo-β-lactamases. Crystal Growth & Design. 2017;17(2):913-924. DOI: 10.1021/acs.cgd.6b01805

[36] Stewart SPD, Kolek SA, Briggs RA, Chayen NE, and PFM Baldock. Random Microseeding: A Theoretical and Practical Exploration of Seed Stability and Seeding Techniques for Successful Protein Crystallization. Cryst. Growth Des. 2011;11(8):3432-3441. https://doi.org/10.1021/cg2001442

[37] Bolanos-Garcia VM, Chayen NE. New directions in conventional methods of protein crystallization. Prog Biophys Mol Biol. 2009 Nov;101(1-3):3-12. doi: 10.1016/j.pbiomolbio.2009.12.006. Epub 2009 Dec 16. PMID: 20018205.

[38] McPherson A, Cudney B. Searching for silver bullets: an alternative strategy for crystallizing macromolecules. J Struct Biol. 2006 Dec;156(3):387-406. doi: 10.1016/j.jsb.2006.09.006. Epub 2006 Oct 11. PMID: 17101277.

[39] Hassell AM, An G, Bledsoe RK, Bynum JM, Carter HL 3rd, Deng SJ, Gampe RT, Grisard TE, Madauss KP, Nolte RT, Rocque WJ, Wang L, Weaver KL, Williams SP, Wisely GB, Xu R, Shewchuk LM. Crystallization of protein-ligand complexes. Acta Crystallogr D Biol Crystallogr. 2007 Jan;63(Pt 1):72-9. doi: 10.1107/S0907444906047020. Epub 2006 Dec 13. PMID: 17164529; PMCID: PMC2483499.

[40] Ito L, Kobayashi T, Shiraki K, Yamaguchi H. Effect of amino acids and amino acid derivatives on crystallization of hemoglobin and ribonuclease A. J Synchrotron Radiat. 2008 May;15(Pt 3):

316-8. doi: 10.1107/S0909049507068598. Epub 2008 Apr 18. PMID: 18421168; PMCID: PMC2394812.

[41] Uyen DT, Huy NT, Trang DT, Nhien NT, Oida T, Hirayama K, Harada S, Kamei K. Effects of amino acids on malarial heme crystallization. Biol Pharm Bull. 2008 Aug;31(8):1483-8. doi: 10.1248/bpb.31.1483. PMID: 18670076.

[42] Manjasetty BA, Turnbull AP, Panjikar S, Büssow K, Chance MR. Automated technologies and novel techniques to accelerate protein crystallography for structural genomics. Proteomics. 2008 Feb;8(4):612-25. doi: 10.1002/pmic.200700687. PMID: 18210369.

[43] Lieberman RL, Culver JA, Entzminger KC, Pai JC, Maynard JA. Crystallization chaperone strategies for membrane proteins. Methods. 2011 Dec;55(4):293-302. doi: 10.1016/j.ymeth.2011.08.004. Epub 2011 Aug 11. PMID: 21854852; PMCID: PMC6186007.

[44] Mir SH, Escher C, Kao WC, Birth D, Wirth C, Hunte C. Generation of recombinant antibody fragments for membrane protein crystallization. Methods Enzymol. 2015;557:201-18. doi: 10.1016/bs.mie.2014.12.029. Epub 2015 Mar 24. PMID: 25950966.

[45] Traenkle B, Kaiser PD, Rothbauer U. Nanobody Platform for Determination of Protein Structure, Application of. Encyclopedia of Analytical Chemistry: Applications, Theory and Instrumentation. Peptides and Proteins. 2016. https://doi.org/10.1002/9780470027318.a9548

[46] Derewenda ZS. Application of protein engineering to enhance crystallizability and improve crystal properties. Acta Crystallogr D Biol Crystallogr. 2010 May;66(Pt 5):604-15. doi: 10.1107/S090744491000644X. Epub 2010 Apr 21. PMID: 20445236; PMCID: PMC3089013.

[47] Manjasetty BA, Büssow K, Panjikar S, Turnbull A. Current methods in structural proteomics and its applications in biological sciences. 3 Biotech. 2012; 2:89-113. https://doi.org/10.1007/s13205-011-0037-1

[48] Cornvik T, Dahlroth SL, Magnusdottir A, Herman MD, Knaust R, Ekberg M, Nordlund P. Colony filtration blot: a new screening method for soluble protein expression in Escherichia coli. Nat Methods. 2005 Jul;2(7):507-9. doi: 10.1038/nmeth767. PMID: 15973420.

[49] An Y, Yumerefendi H, Mas PJ, Chesneau A, Hart DJ. ORF-selector ESPRIT: a second generation library screen for soluble protein expression employing precise open reading frame selection. J Struct Biol. 2011 Aug; 175(2):189-97. doi: 10.1016/j.jsb.2011.04.004. Epub 2011 Apr 16. PMID: 21515383.

[50] Pedelacq JD, Nguyen HB, Cabantous S, Mark BL, Listwan P, Bell C, Friedland N, Lockard M, Faille A, Mourey L, Terwilliger TC, Waldo GS. Experimental mapping of soluble protein domains using a hierarchical approach. Nucleic Acids Res. 2011 Oct;39(18):e125. doi: 10.1093/nar/gkr548. Epub 2011 Jul 19. PMID: 21771856; PMCID: PMC3185438.

[51] Baud F, Karlin S. Measures of residue density in protein structures. Proc Natl Acad Sci U S A. 1999 Oct 26;96(22):12494-9. doi: 10.1073/pnas.96.22.12494. PMID: 10535950; PMCID: PMC22960.

[52] Bolanos-Garcia VM, Chayen NE. New directions in conventional methods of protein crystallization. Prog Biophys Mol Biol. 2009 Nov;101(1-3):3-12. doi: 10.1016/j.pbiomolbio.2009.12.006. Epub 2009 Dec 16. PMID: 20018205.

[53] Banatao DR, Cascio D, Crowley CS, Fleissner MR, Tienson HL, Yeates TO. An approach to crystallizing proteins by synthetic symmetrization. Proc Natl Acad Sci U S A. 2006 Oct 31;103(44): 16230-5. doi: 10.1073/pnas.0607674103. Epub 2006 Oct 18. PMID: 17050682; PMCID: PMC1637565.

[54] Derewenda ZS. The use of recombinant methods and molecular engineering in protein crystallization. Methods. 2004 Nov;34(3):354-63. doi: 10.1016/j.ymeth.2004.03.024. PMID: 15325653.

[55] Waugh DS. Making the most of affinity tags. Trends Biotechnol. 2005 Jun;23(6):316-20. doi: 10.1016/j.tibtech.2005.03.012. PMID: 15922084.

[56] Kobe B, Ve T, Williams SJ. Fusion-protein-assisted protein crystallization. Acta Crystallogr F Struct Biol Commun. 2015 Jul;71(Pt 7):861-9. doi: 10.1107/S2053230X15011061. Epub 2015 Jun 27. PMID: 26144231; PMCID: PMC4498707.

[57] Reddy Chichili VP, Kumar V, Sivaraman J. Linkers in the structural biology of protein-protein interactions. Protein Sci. 2013 Feb;22(2):153-67. doi: 10.1002/pro.2206. Epub 2013 Jan 8. PMID: 23225024; PMCID: PMC3588912.

[58] Yabashi M, Tanaka H. The next ten years of X-ray science. Nature Photon. 2017;11:12-14. https://doi.org/10.1038/nphoton.2016.251

[59] Chapman HN, Fromme P, Barty A, White TA, Kirian RA, Aquila A, Hunter MS, Schulz J, DePonte DP, Weierstall U, Doak RB, Maia FR, Martin AV, Schlichting I, Lomb L, Coppola N, Shoeman RL, Epp SW, Hartmann R, Rolles D, Rudenko A, Foucar L, Kimmel N, Weidenspointner G, Holl P, Liang M, Barthelmess M, Caleman C, Boutet S, Bogan MJ, Krzywinski J, Bostedt C, Bajt S, Gumprecht L, Rudek B, Erk B, Schmidt C, Hömke A, Reich C, Pietschner D, Strüder L, Hauser G, Gorke H, Ullrich J, Herrmann S, Schaller G, Schopper F, Soltau H, Kühnel KU, Messerschmidt M, Bozek JD, Hau-Riege SP, Frank M, Hampton CY, Sierra RG, Starodub D, Williams GJ, Hajdu J, Timneanu N, Seibert MM, Andreasson J, Rocker A, Jönsson O, Svenda M, Stern S, Nass K, Andritschke R, Schröter CD, Krasniqi F, Bott M, Schmidt KE, Wang X, Grotjohann I, Holton JM, Barends TR, Neutze R, Marchesini S, Fromme R, Schorb S, Rupp D, Adolph M, Gorkhover T, Andersson I, Hirsemann H, Potdevin G, Graafsma H, Nilsson B, Spence JC. Femtosecond X-ray protein nanocrystallography. Nature. 2011 Feb 3;470(7332):73-7. doi: 10.1038/nature09750. PMID: 21293373; PMCID: PMC3429598.

[60] Schönherr R, Rudolph JM, Redecke L. Protein crystallization in living cells. Biol Chem. 2018 Jun 27;399(7):751-772. doi: 10.1515/hsz-2018-0158. PMID: 29894295.

[61] Koopmann R, Cupelli K, Redecke L, Nass K, Deponte DP, White TA, Stellato F, Rehders D, Liang M, Andreasson J, Aquila A, Bajt S, Barthelmess M, Barty A, Bogan MJ, Bostedt C, Boutet S, Bozek JD, Caleman C, Coppola N, Davidsson J, Doak RB, Ekeberg T, Epp SW, Erk B, Fleckenstein H, Foucar L, Graafsma H, Gumprecht L, Hajdu J, Hampton CY, Hartmann A, Hartmann R, Hauser G, Hirsemann H, Holl P, Hunter MS, Kassemeyer S, Kirian RA, Lomb L, Maia FR, Kimmel N, Martin AV, Messerschmidt M, Reich C, Rolles D, Rudek B, Rudenko A, Schlichting I, Schulz J, Seibert MM, Shoeman RL, Sierra RG, Soltau H, Stern S, Strüder L, Timneanu N, Ullrich J, Wang X, Weidenspointner G, Weierstall U, Williams GJ, Wunderer CB, Fromme P, Spence JC, Stehle T, Chapman HN, Betzel C, Duszenko M. In vivo protein crystallization opens new routes in structural biology. Nat Methods. 2012

Jan 29;9(3):259-62. doi: 10.1038/nmeth.1859. PMID: 22286384; PMCID: PMC3429599.

[62] Stevenson HP, Lin G, Barnes CO, Sutkeviciute I, Krzysiak T, Weiss SC, Reynolds S, Wu Y, Nagarajan V, Makhov AM, Lawrence R, Lamm E, Clark L, Gardella TJ, Hogue BG, Ogata CM, Ahn J, Gronenborn AM, Conway JF, Vilardaga JP, Cohen AE, Calero G. Transmission electron microscopy for the evaluation and optimization of crystal growth. Acta Crystallogr D Struct Biol. 2016 May;72(Pt 5):603-15. doi: 10.1107/S2059798316001546. Epub 2016 Apr 26. PMID: 27139624; PMCID: PMC4854312.

[63] Stewart PS, Mueller-Dieckmann J. Automation in biological crystallization. Acta Crystallogr F Struct Biol Commun. 2014 Jun;70(Pt 6):686-96. doi: 10.1107/S2053230X14011601. Epub 2014 May 28. PMID: 24915074; PMCID: PMC4051518.

[64] Saridakis E. Perspectives on high-throughput technologies applied to protein crystallization. Protein Pept Lett. 2012 Jul;19(7):778-83. doi: 10.2174/092986612800793154. PMID: 22489783.

[65] Byrappa, K.; Klapper, H.; Ohachi, T.; Fornari, R. Crystal Growth of Technologically Important Electronic Materials; Allied Publishers, Mysore: India, 2003; ISBN 81-7764-375-4.